The M&E TECbook Series

Mammalian Physiology Level II

PHYLLIS A. ALLEN
M.Sc., M.I. Biol.
Senior Biologist, North West Kent College of Technology

Macdonald and Evans

Macdonald & Evans Ltd.
Estover, Plymouth PL6 7PZ

First published 1981

© Macdonald & Evans Ltd. 1981

ISBN: 0 7121 1283 9

Printed in Great Britain by
Hollen Street Press Ltd
Slough

Preface

This book has been written specifically to cover the TEC Mammalian Physiology Level II syllabus with special reference to the generalised mammal, Homo sapiens. It is also useful for students who are studying Biology and Human Biology at "O" and "A" Levels of the G.C.E. examinations and for examinations of the nursing profession.

Each chapter ends with a set of self-assessment questions which, together with cumulative questions after Chapters 5 and 9, enable readers to assess that they have reached the objectives of each chapter and of the book. Assignments to test ability to think through the material studied are also included at the end of each chapter.

Acknowledgment is made to my husband for his help and encouragement during the preparation of this book, to Mr Laurie North, and Dr Edwin Kerr, for their work in editing the text, and to the publishers.

October 1981   P.A.A.

v

Contents

CHAPTER ONE

The Tissues of the Body

CHAPTER OBJECTIVES

After studying this chapter you should be able to:
* identify the various tissues to be found in mammals;
* relate the structure of the tissue to its function in the whole living organism.

INTRODUCTION

This chapter introduces the student to the tissues of the body that have arisen as a result of specialisation required to maintain life above that possible by undifferentiated organisms and will consider how this structure is adapted to a specific function.

The study of the microscopic structure of the tissues of the body is called *histology*. As you will recall from previous study the body of an adult organism, e.g. man, develops from a fertilised egg cell which divides to produce millions of cells which are differentiated to form tissues. A tissue is a group of similar cells specialised in a common way and able to perform a common function.

Tissues are considered in the following four main groups:

(a) epithelial tissues;
(b) muscle tissues;
(c) nervous tissues;
(d) connective tissues.

EPITHELIAL TISSUES

Epithelial tissues are protective tissues which cover surfaces in the body. When they line cavities and organs like the heart, blood vessels and lymphatic vessels they are called *endothelia*.

These tissues always have a small amount of intercellular substance and a basement membrane. A basement membrane is a homogeneous membrane; usually in epithelia it is formed from the surface layer of the underlying connective tissue. It may be formed of flattened connective tissue cells joined together to form a membrane or it may consist of condensed connective tissue ground substance and thus have no cellular structure whatsoever.

To make the study of these tissues easier we can divide epithelia into two types:

(a) simple — one cell layer thick;
(b) compound — more than one layer of cells thick.

Simple epithelia

Squamous epithelium
This tissue (*see* Fig. 1) is composed of very thin and flat cells

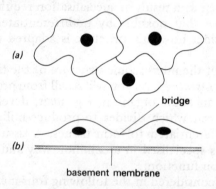

Fig. 1. *Squamous epithelium. (a) Surface view. (b) Side view.*

which fit closely together and are interconnected by cytoplasmic bridges. They are found:

(a) lining the alveoli where the cells form a thin membrane which assists the rapid diffusion of the respiratory gases to and from the blood;
(b) lining the lymphatic and blood vessels where they provide a smooth surface for the passage of liquids;
(c) covering the gut and lining the abdominal cavity as peritoneum which forms lubricating membranes between surfaces that rub together during peristaltic movement.

Cubical epithelium
These cells (*see* Fig. 2) are as wide as they are tall and they line ducts and glands; they are secretory. They are found in:

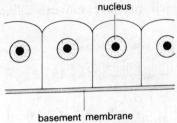

Fig. 2.   *Cubical epithelium.*

(*a*) the uriniferous tubules;
(*b*) the germinal epithelium of the ovary;
(*c*) the thyroid vesicles;
(*d*) the pigmented layer of the retina.

Columnar epithelium
These are tall cells (*see* Fig. 3) with a large nucleus at the base of each cell and a striated border at the free edge. The striated border consists of numerous small, cylindrical projections called microvilli which increase the surface area of the cell. These cells line the gut from the stomach to the anus. Special *goblet cells* produce mucus which protects the cells from the action of the digestive enzymes and helps lubricate the gut as the contents pass through.

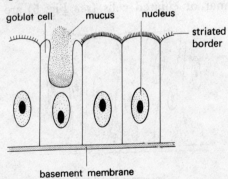

Fig. 3.   *Columnar epithelium.*

Ciliated epithelium

These cells (*see* Fig. 4) are tall and their free edges have whip-like projections of protoplasm called *cilia* which move. Ciliated epithelium resembles columnar epithelium in that it contains goblet cells which produce mucus. Ciliated epithelium is present in:

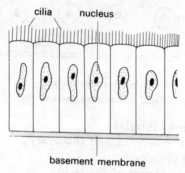

Fig. 4.   *Ciliated epithelium.*

(*a*) the spinal canal where it ensures that the cerebrospinal fluid is kept in motion;

(*b*) the oviducal funnel where the cilia waft the eggs along to the uterus;

(*c*) the trachea and bronchi where dust and bacteria are trapped in the mucus produced by the goblet cells; this is then wafted to the pharynx and swallowed or spat out.

Pseudostratified epithelium

These columnar or ciliated cells (*see* Fig. 5) are always one

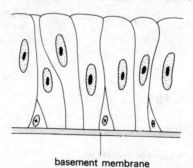

basement membrane

Fig. 5.   *Pseudostratified epithelium.*

cell thick but not all the cells reach the free edge. They are found in the olfactory mucosa where they protect the underlying cells from the cold, dry, dirt-laden air which is inhaled during breathing.

Compound epithelia

These tissues are found in the more exposed parts of the body. They consist of several layers of cells and are therefore more robust than simple epithelia and give greater protection.

Stratified epithelium

This consists of a number of layers of cells (*see* Fig. 6) which arise from a germinative layer of cubical cells lying on a basement membrane. The germinative layer is constantly dividing by *mitosis* and the cells are being pushed upwards by new cells as they are formed. As these cells approach the surface they become flattened and are eventually lost from the outer surface.

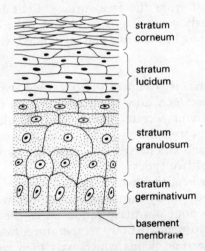

stratum corneum

stratum lucidum

stratum granulosum

stratum germinativum

basement membrane

Fig. 6. *Stratified epithelium.*

Stratified epithelium is found in the skin, buccal cavity, the oesophagus and in the vagina; in all these situations it provides protection to the body against wear and tear. In skin the outer cells become cornified to form a tough, non-living waterproof layer.

Transitional epithelium

This consists of three or four layers of small cells (*see* Fig. 7). Transitional epithelium is capable of considerable distention and is found in the urinary bladder, the pelvis of the kidney and in the ureter.

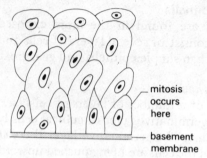

mitosis
occurs
here

basement
membrane

Fig. 7. *Transitional epithelium.*

Mitosis is frequent in the lowest layer of cells and these replace any lost from the free surface. Cells are, of course, lost continuously as they grow old and flake off. They are then carried along in the urine and are voided with it.

Glands

Epithelial tissue is frequently concerned with secretion; already we have seen how the goblet cells in columnar and ciliated epithelium secrete mucus. Multicellular *glands* are formed of secretory epithelial cells which are bound, together with blood vessels and nerves, by connective tissue.

Glands are of two types:

(a) the *exocrine* glands which produce their secretion on to a surface directly or by means of a duct, e.g. sweat glands in the skin which pour out perspiration on to the body surface and thus help to regulate body temperature, (*see* Fig. 8); and

(b) the *endocrine* glands which pass their secretion, hormones, directly into the blood stream. Examples of these are the thyroid gland which produces thyroxine and the islets of Langerhans in the pancreas which produce insulin.

MUSCLE TISSUES

These are contractile tissues which do not secrete intercellular substances. They consist of elongated cells or fibres. These fibres are held together by areolar connective tissue. The fibres contain highly specialised cytoplasm called *sarcoplasm* which contains the myofibrils; these are the contractile fibrils of muscle tissue.

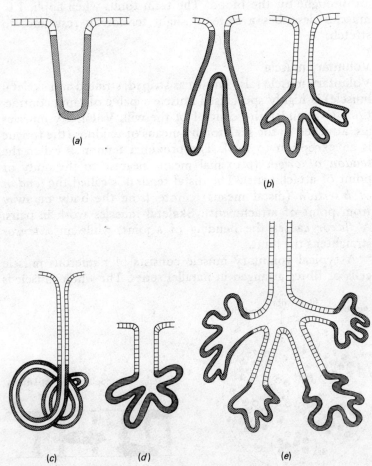

Fig. 8. *(a) Simple tubular gland as in the <u>intestinal glands</u>.*
(b) Simple branched tubular glands as in the <u>gastric glands</u>.
(c) Simple coiled tubular gland as in the <u>sweat glands</u>.
(d) Simple branched alveolar gland as found in the sebaceous glands.
(e) Compound gland as in the salivary and mammary glands.

NOTE: Secretory regions shown as darker areas.

Muscular contraction is due to a change in the arrangement of the protein molecules of which the muscle is composed, and may be short and sudden or slow or rhythmical. It is brought about by energy obtained from the oxidation of glucose which occurs in tissue respiration. The necessary food and oxygen are brought by the blood. The term tonus when applied to muscles describes a state of slight tension or resistance to stretch.

Voluntary muscle

Voluntary muscle (also known as striped, striated and skeletal muscle) is highly specialised muscle capable of rapid contraction. It is under the control of the will. Voluntary muscles are attached to the skeleton by means of tendons (the tongue is an exception to this). The proximal tendon is called the *tendon of origin* (proximal means nearest to the body or point of attachment). The distal tendon is called the *tendon of insertion* (distal means remote from the body or away from point of attachment). Skeletal muscles work in pairs. A *flexor* causes the bending of a joint, while an *extensor* straightens the joint.

A typical voluntary muscle consists of numerous muscle cells or fibres arranged in parallel series. The whole muscle is

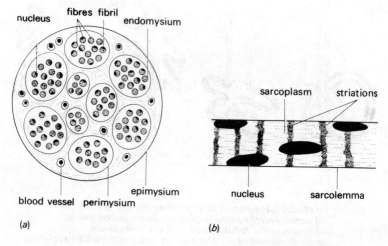

Fig. 9.   *(a) Section (T.S.) through a whole voluntary muscle.*
          (b) Fibre.

enclosed in a sheath of connective tissue called the *epimysium*. Ingrowths of the sheath divide the muscle into bundles or *fasciculi*. The sheath of each bundle is called the *perimysium* and ingrowths of it separate and cover individual fibres forming their cover, the *endomysium*. Inside the endomysium each fibre has an elastic membrane called the *sarcolemma*. The coenocytic fibres have their elongated nuclei situated just below the sarcolemma (*see* Fig. 9).

Each fibre is composed of many *myofibrils* lying in the sarcoplasm and the myofibrils contain the proteins *myosin* and *actin* which are linked together by bridges and appear as alternating light and dark bands. These bands give the voluntary muscle its characteristic stripes. The finely branched motor nerves which innervate the voluntary muscle tissue run obliquely, or at right angles, to the fibres and terminate in end plates whose fine fibrils penetrate the sarcolemma (*see* Fig. 10).

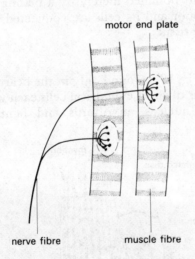

Fig. 10. *Motor end plates.*

Involuntary muscle

This muscle (*see* Fig. 11 — also known as unstriped, unstriated, splanchnic or visceral muscle) occurs in the walls of the gut where it is responsible for peristaltic action. It is also found in the walls of blood vessels, bladder, ducts, the dermis of the skin, the iris and ciliary body of the eye. The muscle action is a simple, involuntary movement, i.e. it is not under the control of the will.

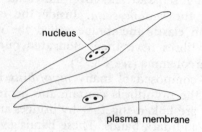

Fig. 11. *Involuntary muscle.*

The muscle is innervated by the autonomic nervous system (*see* Nervous tissues). Its fibres are long and spindle shaped with no sarcolemma but bounded merely by a plasma membrane. Blood supply is poor and its cells are connected together by yellow connective tissue.

Cardiac muscle

This muscle (*see* Fig. 12) is found only in the heart. The fibres consist of columns of short cylindrical cells each with its own centrally placed nucleus, myofibrils and faint transverse

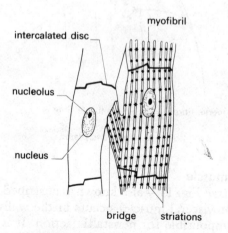

Fig. 12. *Cardiac muscle.*

striations. Adjoining fibres are linked by means of cross connections. No true sarcolemma is present, but the cells are rich in sarcoplasm. The tissue is richly supplied with blood vessels and is innervated by the vagus nerve and the autonomic system. Cardiac muscle cells are joined end to end at specialised zones called intercalated discs.

NERVOUS TISSUES

Nervous tissue forms the communication system between the different parts of the body. It is highly specialised in its function. The central nervous system consists of the brain and spinal cord. Twelve pairs of cranial nerves carry messages to and from the brain, and paired spinal nerves carry messages to and from the spinal cord. The autonomic nervous system, consisting of the parasympathetic and sympathetic system, also contributes to the co-ordination of the body as a whole. Sense organs, such as the eye, ear, small receptors, taste buds, heat receptors and pressure receptors make the body sensitive to changes in the external environment. The endocrine organs, which we considered in the section on epithelial tissues, help to maintain a constant internal environment within the body.

Obviously with such diversity of function there is great diversity of structure. In order that we may understand the basic histological structure of nervous tissues we will consider them under several headings.

Neurones

The basic unit of the nervous tissue is the nerve cell or neurone. This is composed of a cell body with cell processes, one of which is always either a nerve fibre or else the axis cylinder of a nerve fibre. It is called the *neuraxon* and its *end processes* carry impulses from the cell body. The other processes carry impulses into the cell body and are called *dendrons* because their branches resemble those of a tree; dendrons give rise to smaller branches called dendrites. The terminal processes of the dendrons and axons end in synaptic knobs which are close to another neurone or effector cells. The slight gaps that occur between the knobs of one nerve and another, or between a nerve cell and an effector cell, are called *synapses*.

The cell bodies vary in shape, e.g. round, pyramidal, stellate and flask shaped. The nuclei are always large with nucleoli. Stained preparations show the presence of Nissl granules except in the region of the axon. Nissl granules are clumps of ribo nuclear proteins probably concerned with synthesis of substances within the nerve cell.

Nerve cells are found either centrally within the grey matter of the brain and spinal cord or peripherally in the sense organs or ganglia (*see* Fig. 13).

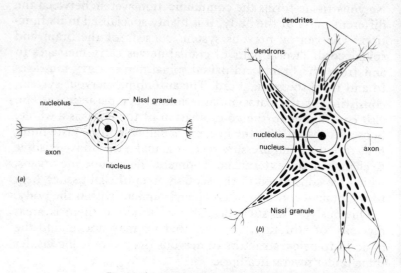

Fig. 13. *(a) Bipolar neurone. (b) Neurone.*
NOTE: *A neurone with one process which branches beyond the cell body into an axon and a dendrite is termed unipolar. A neurone with two processes is bipolar and with more than two is multipolar.*

Neuroglia

Between the nerve cells and nerve fibres there are other cells called *neuroglia*. Their name is derived from the Greek and it means "nerve glue". These neuroglia cells develop from the ciliated epithelium that lines the cavities of the brain and spinal cord and they form a sponge-like packing tissue between neurones in those organs.

Nerve fibres

These are the axons of nerve cells and there are two kinds.

Medullated, or myelinated, nerve fibres

These are so named because they have a fatty, insulating myelin sheath on the outside of the axon or axis cylinder. This is "pinched in" at intervals thus forming the *nodes of Ranvier*. Outside the medullary sheath there is a tough, outer coat of nucleated cells called the *neurilemma*. There is usually one nucleus for each internode, i.e. the area between two nodes of Ranvier (*see* Fig. 14).

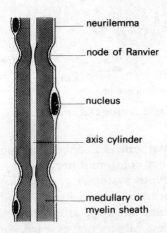

Fig. 14. *Myelinated nerve fibre.*

The white matter of the brain and spinal cord is composed chiefly of medullated fibres as are the cerebrospinal nerves.

Amyelinated or non-medullated nerve fibres

There are found mainly in the *autonomic nervous system*. These fibres are devoid of a myelin sheath and simply consist of a neuraxon enclosed by a neurilemma and connective tissue.

Nerves

Nerves consist of bundles of nerve fibres which are bound together by connective tissue called the *epineurium*. Each bundle is covered by a connective tissue sheath called the *perineurium*, ingrowths of which run between the individual fibres as the *endoneuria*. Within the connective tissue there are blood vessels, lymphatics and some adipose tissue (*see* Fig. 15).

Nerves are of three different types; namely motor, sensory and mixed. Motor or efferent nerves carry motor impulses out from the central nervous system to voluntary muscles,

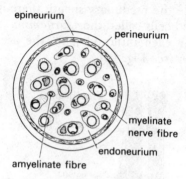

Fig. 15. *Section (T.S.) through a nerve.*

e.g. oculomotor, patheticus, abducens (all muscles which move the eyeball). Sensory or afferent nerves carry sensory impulses to the central nervous system, e.g. olfactory (concerned with smell), optic (concerned with sight), and auditory (concerned with hearing). Some sensory, afferent nerves contain autonomic nerve fibres. Mixed nerves carry both afferent and efferent fibres, e.g. trigeminal, facial, glossopharyngeal, vagus and all spinal nerves.

Spinal cord

The spinal cord is the thick nerve cord which is protected by the vertebral column. It is surrounded by the *dura mater*, a tough membranous coat, behind this is a non-vascular *arachnoid* and surrounding the nervous tissue there is the fibrous *pia mater* with its rich blood supply. Mid-dorsally there is a vein, ventrally there is an artery, while dorsally and ventrally a fissure runs into the cord. If the spinal cord is sectioned and stained a transverse section will show that there is a broad H-shaped mass of grey matter consisting of neurons, with, centrally, a canal lined by ciliated epithelium containing cerebrospinal fluid. The four limbs of the grey matter are called right and left dorsal horns and right and left ventral horns (*see* Fig. 16).

The peripheral white matter is composed of nerve fibres, connective tissue strands and neuroglia, e.g. packing tissue cells.

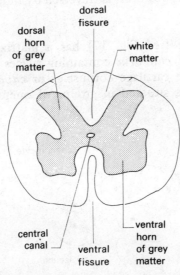

Fig. 16. *Section (T.S.) through the spinal cord.*

The brain
The position of the grey and white matter is reversed, i.e. the grey matter is peripheral and white matter lines the central cavity or ventricle.

CONNECTIVE TISSUES

These tissues make up a large part of the body. They conduct, bind and support the parts of the body and have an intercellular substance called a *matrix* which is fluid or semi-fluid.

Skeletal connective tissues
These consist of bone and cartilage. They form a basic structure for the attachment of the muscles to act as a series of levers, e.g. in the arms and legs. The bulk of the framework in the adult is composed of hard, rigid bone which supports and protects the vital organs. The skeleton of the embryo is composed of cartilage which is replaced by bone as the embryo

develops. Cartilage, unlike bone, has no blood vessels, but its matrix is permeable to blood plasma which carries nutrients to the cells. Bone, however, is traversed by blood vessels.

Hyaline cartilage

Hyaline cartilage (*see* Fig. 17) has a matrix composed of clear translucent *chondrin* containing spaces called *lacunae* which house the cartilage corpuscles or *chondrocytes*. The chondrocytes are situated in groups of one, two, four and eight.

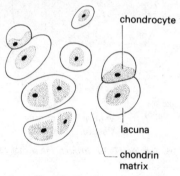

Fig. 17. *Hyaline cartilage.*

Cartilage is covered by a tough protective fibrous membrane called the *perichondrium*. Besides being found in the skeletons of mammalian embryos, cartilage is found at the ends of long bones of adults where it forms articular cartilage. Here its resilience prevents the bones grating together. It also is present in the xiphisternum, the suprascapular cartilage, the ends of the ribs and the thin plates between the diaphyses and epiphyses of the long bones (*see* Bone: page 17).

White fibrous cartilage

This cartilage (*see* Fig. 18) has its matrix packed with bundles of white fibres which add firmness to the flexibility possessed by the cartilage. It is found in the intervertebral discs where it cushions the vertebrae against shocks and in the patella where it protects the knee joint. It is also found in the pubic symphysis.

Yellow elastic cartilage

This cartilage (*see* Fig. 19) contains yellow, branching, anastomosing fibres in its matrix. These elastic fibres enable recovery of shape after distortion. It is found in the pinna of the ear, the epiglottis, the Eustachian tube and the external auditory meatus.

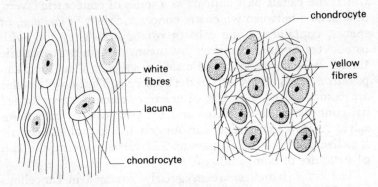

Fig. 18. *White fibrous cartilage.*   Fig. 19. *Yellow elastic cartilage.*

Bone

Bone (*see* Fig. 20) is a hard rigid tissue with calcareous salts deposited in its gelatinous matrix. There are two types of bone, namely *spongy bone*, which is also called *cancellous bone*, and *compact*, or *dense*, *bone*. If a long bone is cut in half longitudinally its structure can be examined: the two ends are called the epiphyses and the shaft between them is the diaphysis; the epiphysis consists of a spongy arrangement of anastomosing bony spicules with marrow between the meshes. Its outer covering is compact bone. Cancellous or spongy bone forms a narrow strip around the central marrow cavity of the diaphysis but the main part of the shaft outside this is compact bone. The structure of cancellous and compact bone is essentially the same but in cancellous bone the spaces are large whilst in compact bone the canals are small.

Skull bones consist of a sandwich of two plates of compact bone with cancellous bone between them.

The short bones are usually composed of cancellous bone with a covering of compact bone on the outside.

Compact bone has a series of roundish *Haversian canals* through which blood vessels carrying nutrients and nerves pass. Communication with these canals are the *Volkmann's canals* which pierce the bone from the outer and inner surfaces. Bone therefore has a complete system of channels containing blood vessels and nerves. The bony matrix through which the canals pass appears as a series of concentric layers, or *lamellae*, between which are concentric rings of *lacunae*, or spaces, containing bone cells or *osteocytes*. They link with osteocytes of adjacent rings by means of thin protoplasmic threads which run in canals called *canaliculi*. Thus food can pass from the blood vessels to the bone cells. The bony lamellae are composed of a matrix of gelatine which is strengthened by connective tissue fibres and by a deposit of calcareous salts. A Haversian canal surrounded by several concentric lamellae and bone cells makes up a Haversian system, the unit of structure of compact bone.

The bone lamellae are irregularly arranged in cancellous bone and there are no Haversian systems.

The vascular connective tissue sheath covering bone is called the *periosteum*. It is concerned with the nourishment of the tissue and brings blood vessels and nerves to it. It is also concerned with the attachment of muscles and tendons.

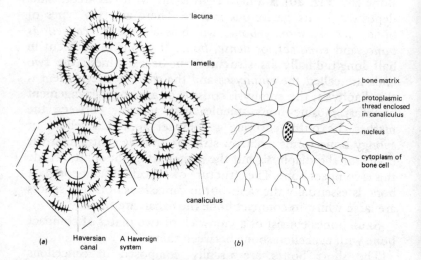

Fig. 20. *(a) Section (T.S.) through bone. (b) A bone cell in its lacuna.*

Other connective tissues

Areolar connective tissue

This is packing tissue found surrounding organs (*see* Fig. 21).
It connects skin to the rest of the body, peritoneum to the
body wall and to the muscle of the gut. It connects two layers
of squamous epithelium forming the mesenteries and it surrounds nerves and blood vessels.

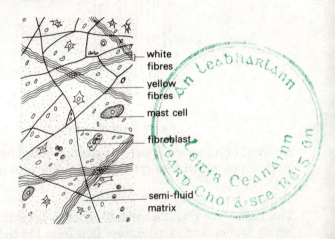

Fig. 21. *Areolar connective tissue.*

It consists of fine, branched, yellow elastic fibres and
bundles of coarse wavy, unbranched, white inelastic fibres
crossing each other to form a mesh. The ground tissue is composed of gelatin containing cells of various types. Fibroblasts
secrete the fibres, oval mast cells produce the matrix and there
are also migratory, amoeboid leucocytes. The tissue is capable
of considerable stretch and recovery. The limit of the stretch
is the straightening of the wavy inelastic fibres. Recovery
depends on the elasticity of the yellow elastic fibres.

White fibrous tissue

This tissue (*see* Fig. 22) consists of closely packed white
collagen fibres arranged in parallel series with rows of fibroblasts between them. The fibre bundles are bound together
with areolar connective tissue.

The white fibres make this tissue very strong with limited flexibility. It covers the kidney, forms the sclerotic coat of the eye and it forms the transparent tissue of the cornea. The *perichondrium* which covers cartilage and the *periosteum*

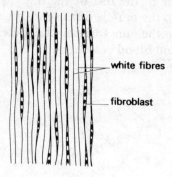

Fig. 22. *White fibrous tissue.*

covering bone are composed of strong white fibrous tissue, as are tendons which connect voluntary muscle to bone.

Yellow elastic tissue

This tissue (*see* Fig. 23) consists of parallel elastic yellow fibres which branch and anastomose. It is found in the ligamentum nuchae where its strength and elasticity is used to attach the skull to the cervical vertebrae. It is also present in the walls of arteries and in the bronchioles.

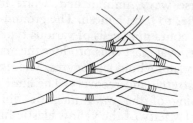

Fig. 23. *Yellow elastic tissue.*

Adipose tissue

This tissue (*see* Fig. 24) has the appearance of plant parenchyma. It is found mixed with other tissues, e.g. around the kidney and heart and under the skin. The cells contain fat droplets which push the nucleus to one side. When permanent slides are made the fat dissolves and the cell appears empty except for the nucleus.

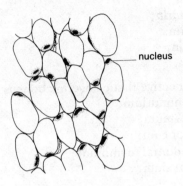

Fig. 24. *Adipose tissue.*

CONCLUSION

You have now finished your study of the tissues of the body. There is one connective tissue which has been omitted, that of the very important, specialised tissue, blood. This vital connective tissue will be dealt with in the next chapter. Before proceeding to that here are some questions to enable you to assess how much you have learned so far.

SELF-ASSESSMENT QUESTIONS

1. The lining of the gut is composed of:
   (a) squamous epithelium;
   (b) ciliated epithelium;
   (c) endothelium;
   (d) adipose tissue;
   (e) columnar epithelium.

2. Which tissue lines the trachea?
   (a) squamous epithelium;
   (b) cubical epithelium;

    (c) ciliated epithelium;
    (d) columnar epithelium;
    (e) stratified epithelium.

3. The flexible matrix of cartilage is composed of:
    (a) chitin;
    (b) chondrin;
    (c) calcium;
    (d) elastin;
    (e) fibrin.

4. The connective tissue covering bone is called:
    (a) perichondrium;
    (b) periosteum;
    (c) peritoneum;
    (d) periodontal membrane;
    (e) perimysium.

5. A tendon:
    (a) joins muscle to bone;
    (b) holds bones in place at a joint;
    (c) fills the gaps between the other tissues in organs;
    (d) protects soft underlying tissue;
    (e) covers the outside of organs.

6. Mixed nerves contain:
    (a) afferent fibres;
    (b) afferent and efferent fibres;
    (c) efferent fibres;
    (d) autonomic nerve fibres;
    (e) neurones.

7. Collagen fibres are flexible, very strong and inextensible.
They are found in:
    (a) areolar connective tissue;
    (b) hyaline cartilage;
    (c) bone;
    (d) cubical epithelium;
    (e) striated muscle.

8. Elastic fibres are found in the cartilage of:
    (a) the intervertebral discs;
    (b) the nose and the pinna;

(c) the skeleton of the embryo;
(d) the ends of the long bones;
(e) the ends of the ribs.

9. The covering or sheath of the axon fibre of a neurone consists of:
(a) white fibres;
(b) myelin;
(c) amylose;
(d) perimysium;
(e) yellow fibres.

10. Which of the following tissues can change shape and contract?
(a) cartilage;
(b) cardiac muscle;
(c) ligaments;
(d) squamous epithelium;
(e) tendons.

ASSIGNMENTS

1. Make a large chart showing all the tissues of the body. Divide the page up into three columns and put the type of tissue in the first one; where it is found in the second one; and the function of the type of tissue in the last column.

2. Show how the structure of epithelial tissues is related to their function.

CHAPTER TWO

The Blood

```
CHAPTER OBJECTIVES

After studying this chapter you should be able to:
* describe the structure of blood;
* identify the main types of blood cells;
* enumerate the functions of the blood;
* state the functions carried out by each cell type;
* give an account of the composition of plasma;
* explain the role of haemoglobin in transporting oxygen
  from lung to tissues;
* show how carbon dioxide is transported in blood and
  lymph.
```

INTRODUCTION

This chapter considers the chemical and physical composition
of blood and relates its structure to the functions it carries out.

Blood is a connective tissue with a large amount of fluid
matrix called plasma in which *erythrocytes,* the red blood
corpuscles,*leucocytes,* the white corpuscles and *thrombocytes,*
or blood platelets, float. Plasma is a straw-coloured liquid of
which 90 per cent is water, and is discussed in detail on page
51.

ERYTHROCYTES

Human erythrocytes are biconcave non-nucleated disc-shaped
cells. Each one is 8.5 μm in diameter, 2 μm thick, but only
1 μm thick in the centre. An erythrocyte is bounded by a
flexible envelope which encloses haemoglobin and inorganic
ions such as sodium, potassium, calcium, magnesium, chloride
and phosphate. Erythrocytes are produced in the red bone
marrow of the sternum, skull bones, vertebrae and the ends
of long bones in the adult but in the intra-uterine life they
are produced in the yolk sac, spleen, liver and lymph nodes.
There are approximately 5 000 000 erythrocytes per mm^3 in
the blood of a man and 4 500 000 per mm^3 in a woman. They

carry oxygen combined with the red pigment haemoglobin and also some carbon dioxide. The erythrocytes remain in the circulating fluid for about 120 days and at the end of their life are phagocytosed by cells in the liver, spleen and marrow.

Destruction of erythrocytes

Erythrocytes are phagocytosed by macrophages in the liver, bone marrow and spleen (*see* page 29). The haem part is split from the globin part which is degraded to its constituent amino acids. The haem part is converted into biliverdin which is further changed into bilirubin. The iron is removed and resynthesised to form new haemoglobin. Bilirubin combines with albumin and is absorbed by the liver cells, where it is used to form bile salts, then it passes into the small intestine to emulsify fats. Finally, bacteria act upon it converting it to urobilinogen. This is the substance which give faeces their brown colour. Some urobilinogen is absorbed into the blood from the intestinal walls, taken into the kidneys and filtered out, which gives urine its yellow colour. Some diseases, e.g. sickle cell anaemia, result in excessive breakdown of erythrocytes; too much bilirubin is produced and is released into the blood stream in large excess resulting in jaundice, when the sclerotic coat of the eye, the mucous membranes and skin turn yellow. Infective hepatitis causes a similar effect because the liver cannot take in pigment effectively. Gall-stones can block the bile duct and this results in bilirubin being distributed around the body, creating yellowing of the skin.

GASEOUS TRANSPORT

How oxygen is carried in the blood

Air taken into the lungs during inspiration is a mixture of about 4/5 nitrogen, 1/5 oxygen and a small amount of carbon dioxide. The concentration of gases in the body is expressed in terms of their partial pressures measured in pascals (1 pascal = 1 N/m^2). Air passes into the respiratory system and mixes with air which has not been completely exhaled in the last expiration. This air contains less oxygen and more carbon dioxide than the incoming air so that the air which eventually reaches the alveoli has a slightly lower oxygen content and a slightly higher carbon dioxide content than the air outside the body.

Blood from respiring tissues, passing through the capillaries surrounding the alveoli, has a lower partial pressure of oxygen and a higher partial pressure of carbon dioxide than the air present in the alveoli. There is, therefore, a concentration gradient which favours oxygen diffusing into the blood from the alveoli and carbon dioxide diffusing out from the blood into the alveoli. When blood leaves the alveolus it has the same partial pressure of oxygen and carbon dioxide as the air in the alveolus, i.e. slightly lower oxygen content and slightly higher carbon dioxide content than the air outside the body.

To reach the haemoglobin in the erythrocyte the oxygen has diffused from the cavity of the alveolus, through its thin wall composed of squamous epithelium, through the squamous epithelium wall of the capillary and through the thin wall of the erythrocyte. Here it combines with the haemoglobin which has a great affinity for oxygen to form oxyhaemoglobin. Each haemoglobin molecule is composed of four iron-containing haem groups which can each combine with a molecule of oxygen. The combination is loose and takes place readily and easily. Oxygen is taken up by the erythrocytes in the lungs and given up to the respiring tissues while the erythrocytes are circulating through the body. The oxygen splits off from the haemoglobin leaving reduced haemoglobin which returns to the lungs in the blood to repeat the process.

In the tissues rapid respiration causes a continuous production of carbon dioxide which diffuses from the high concentration of the respiring cells into the low concentration of the blood. This is the reverse of what happens in the lungs, where carbon dioxide is released continuously into the lumen of the alveolus and there is a continuous fresh supply of oxygen and a low concentration of carbon dioxide because the intercostal muscles and diaphragm work together to bring about inspiration and expiration.

How carbon dioxide is carried in the blood

When carbon dioxide is produced in the tissues, as a result of the oxidation of food, it diffuses into the blood stream; part dissolves in the plasma water, part dissolves in the water present in the erythrocytes and the rest attaches itself to the globin part of the haemoglobin molecule in the form of *carbaminohaemoglobin*.

In the erythrocyte, the enzyme carbonic anhydrase speeds the conversion of carbon dioxide and water into carbonic acid. This acid dissociates into hydrogen ions bearing a positive charge and bicarbonate ions bearing a negative charge.

The bicarbonate ion concentration increases inside the erythrocyte until the bicarbonate ions begin to diffuse out; this upsets the electrical balance of the erythrocyte and so negatively-charged chloride ions from the plasma move into the cell to restore the electrical neutrality; this is known as the *chloride shift*.

The hydrogen ions produced from the dissociation of the carbonic acid could lower the pH of the cell appreciably but the haemoglobin in the erythrocyte prevents this, i.e. acts as a buffer. In the plasma, proteins buffer the hydrogen ions by forming weak proteinic acids. Carbon dioxide entering the erythrocyte from respiring cells displaces the oxygen from its loose connection with the haemoglobin, making it freely available to diffuse out into the tissues and oxidise more food.

In the lungs, blood laden with carbon dioxide is present in the capillaries surrounding the alveoli. The enzyme carbonic anhydrase catalyses the reaction which liberates carbon dioxide from the carbonic acid in the erythrocyte, utilising hydrogen ions from the erythrocyte and bicarbonate ions from the plasma. The latter move back into the erythrocyte, once again upsetting the electrical balance of the cell. This time, an excess of negative particles inside the erythrocyte is brought about. This is rectified by chloride ions shifting back into the plasma.

The high concentration of oxygen in the alveolar air favours the diffusion of oxygen from the alveoli into the erythrocytes where it displaces carbon dioxide from the haemoglobin and then combines with it producing oxyhaemoglobin. The carbon dioxide is released, diffuses into the alveoli from the blood and is expired as a result of the expiratory movements of the diaphragm and the chest wall.

Haemoglobin and carbon monoxide

Haemoglobin takes up carbon monoxide much more readily than it takes up oxygen. This reaction results in the formation of the very stable compound carboxyhaemoglobin. If this continues all the haemoglobin is converted into carboxyhaemoglobin, the blood loses its ability to carry oxygen and death occurs.

LEUCOCYTES

Leucocytes or white blood corpuscles are nucleated amoeboid cells; unlike erythrocytes they do not contain any pigment. Human blood contains approximately 7000–8000 leucocytes per mm^3. This white cell (leucocyte) count varies throughout the day; it is lowest when the body is at rest and generally increases with exercise, chiefly due to an increase in polymorphs (*see* below).

Leucocytes are colourless but their features can be revealed using Wright's stain. They are classified into five types on the basis of their reaction to staining techniques.

Polymorphs or neutrophils

These make up 40–75 per cent of the total number of leucocytes. They are large and have finely granular cytoplasm with irregular or fragmented nuclei. Both nucleus and cytoplasm stain blue. Polymorphs are formed in the bone marrow.

Functions
The granules in the cytoplasm contain peroxidase enzymes. These cells are active at sites of inflammation where they phagocytose (ingest and destroy by enzymic action) bacteria. They liquefy and absorb injured and dead cells and speed healing processes.

Lymphocytes

These make up 20–45 per cent of the leucocyte population. The majority of them are about the same size as an erythrocyte but some are smaller. The cells are non-granular; the nucleus stains deep blue, the cytoplasm stains pale blue. Lymphocytes are formed mainly in the lymphatic tissue but some are produced in the red bone marrow.

Functions
These slightly amoeboid cells pass through the walls of vessels; a process called *diapedesis*. They travel through tissues; during digestion large numbers of lymphocytes pass through the walls of the alimentary canal. They are slightly phagocytic and are concerned with the formation of globulin. Small lymphocytes play a major role in the defensive mechanism of the body (immune response) against infection by bacteria, viruses and the toxins they produce. Bacteria, viruses and their toxins are classed as antigens and lymphocytes produce antibodies in response to their presence. The antibody and antigen combine to form a harmless complex which is then phagoctyosed by large lymphocytes, called macrophages.

Eosinophils or acidophils
These make up 1—6 per cent of the leucocytes. They are about the same size or slightly larger than the polymorphs. The cytoplasm stains faint blue and has many coarse granules which stain red with eosin. The nucleus is horseshoe-shaped or lobed. Eosinophils are formed in the bone marrow.

Function
The cells are actively amoeboid. These cells increase considerably in cases of parasitic worm infection and certain allergic conditions; probably these cells have a detoxifying role (the removal of foreign substances from the body).

Monocytes
These are the largest type of leucocyte, being from 15—20 μm diameter. The cytoplasm contains no granules and stains blue, the nucleus is round or oval and it too stains blue. Monocytes make up 2—10 per cent of the leucocytes. Monocytes are formed in the lymphatic system and probably also in the spleen.

Functions
They are amoeboid, carry out phagocytosis and produce globulin.

Basophils
These are present as 1 per cent of the leucocytes in the child;

they are rare in adults. The cytoplasm, which has large granules, stains blue; the nucleus is bilobed, sometimes kidney-shaped. Basophils are formed in the bone marrow.

Function
They appear to control the viscosity of connective tissue.

THROMBOCYTES

Thrombocytes or blood platelets are small, round discs approximately 2–3 μm diameter. They are granular, proto-plasmic fragments probably mitochondrial in nature. In man there are approximately 150,000–400,000 per mm^3 of blood. Thrombocytes are formed in the bone marrow and aid in the clotting of blood.

Figure 25 gives a summary of the different types of blood cells.

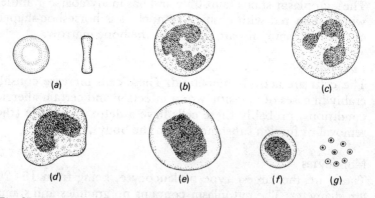

Fig. 25. *Blood cells. (a) Erythrocyte, with side view: approximately 5 000 000/ mm^3, originates in bone marrow, no nucleus, cytoplasm contains haemoglobin. (b)–(f) Leucocytes: approximately 7000/mm^3, have nuclei. (b) Polymorph. (c) Eosinophil. (d) Basophil: granular cytoplasm, originates in bone marrow. (e) Monocyte. (f) Lymphocyte: non-granular cytoplasm, originates in the lymphatic system. (g) Thrombocyte.*

CLOTTING OR COAGULATION OF BLOOD

This is a complex process involving at least thirteen steps, of which the main ones are outlined here.

Plasma contains soluble fibrinogen. When clotting occurs this is converted into insoluble fibrin. Also two antagonistic substances are present in blood; antithrombin which prevents clotting and prothrombin which in the presence of calcium ions forms thrombin.

When a blood vessel is damaged, the endothelial cells lining the walls and injured platelets in the blood release the enzyme *thrombokinase* which neutralises the antithrombin. The prothrombin is converted into thrombin which causes the fibrinogen to be deposited as long threads of fibrin. These form a mesh in which corpuscles are deposited, forming a clot.

FUNCTIONS OF THE BLOOD

Human blood makes up approximately 7 per cent of the body weight in an adult; there are approximately 5.5 litres in volume. It is a specialised tissue having many functions.

Transport system

Blood is a transport system and carries:

(a) oxygen from the lungs to the tissues where it is required;

(b) carbon dioxide from the tissues where it is produced to the lungs where it is expired;

(c) waste materials from the cells which produce them to the organs that get rid of them, e.g. amino acids in excess of the body's requirements and amino acids produced as a result of the breakdown of cells are taken to the liver where they are converted into urea which is then carried in the blood to the kidneys where it is eliminated in the form of urine;

(d) hormones from the endocrine organs where they are produced, to the body areas which are chemically influenced by them, e.g. insulin produced by the islets of Langerhans in the pancreas is carried in the blood to the liver where it affects the conversion of glucose into animal starch or glycogen;

(e) enzymes, some of which, e.g. carbonic anhydrase, are required for use within the blood and others which are required for cell use.

Water content

Blood regulates the water content of the body. The hydro-static and osmotic pressures of the blood influence the water content of the other body fluid compartments.

pH

Blood regulates the pH of the body by means of its buffering systems.

Temperature

Blood regulates the temperature of the body. If the body is too hot the capillaries in the skin dilate and bring more blood to the surface of the skin for cooling. Since blood moves around the body, it tends to equalise the temperature of different parts. It also regulates temperature in that it brings oxygen to the tissues together with food which the cells can use to produce their own heat energy.

Protection

Blood protects the body by clotting if there is wounding; the clotting prevents blood loss. It also protects the body by means of antibodies, antitoxins and its phagocytes which engulf bacteria.

BLOOD GROUPS

Human blood can be classified into four groups, A, B, AB and O, according to the type of antigen present in the person's erythrocytes (a protein which incites the formation of an antibody is an antigen).

Antibodies (agglutinogens) are carried in the plasma and they are represented by the symbols a, b, ab and o. If blood form incompatible individuals is mixed, the erythrocytes clump together, or agglutinate, blocking the capillaries. Later lysis occurs and the cells burst, liberating haemoglobin into the plasma. This sequence of events occurs if antigens in the erythrocytes are complementary to antibodies in the plasma, i.e. A-antigen on red cells will clump with anti-A(a) agglutinogen in the serum. B-antigen on red cells will clump with anti-B(b) agglutinogen in the serum. Anti-A serum has no effect on blood group B cells and anti-B serum has no effect on blood group A cells.

(a) Group A. The blood group A is found in about 40 per cent of the population. Group A blood has A antigens in the erythrocytes and b antibodies in the plasma.

(b) Group B. About 10 per cent of the population has blood group B which contains B antigens in the erythrocytes and a antibodies in the plasma.

(c) Group AB. Group AB blood is only found in about 5 per cent of the population. The erythrocytes contain A and B antigens but neither a nor b antibodies are present in the plasma.

(d) Group O. Group O blood is found in about 45 per cent of the population and has neither A nor B antigens in the erythrocytes but both a and b antibodies in the plasma.

Safe transfusion of blood between individuals can only occur provided the recipient's blood does not contain antibodies in the plasma which correspond to the antigens in the erythrocytes of the donor, otherwise agglutination will occur. It does not matter if the antibodies present in the plasma of the donor are incompatible with the antigens present in the erythrocytes of the recipient because the dilution of the blood minimises the agglutination effect. Since agglutination can cause the death of the patient it is important that cross-matching is carried out carefully.

Reactions occuring when blood of different groups are mixed can be summarised as follows, where na stands for non-agglutination and A stands for agglutination.

TABLE I. MIXING OF BLOOD GROUPS.

Donor	Recipient			
	Oab	Ab	Ba	ABo
Oab	na	na	na	na
Ab	A	na	A	na
Ba	A	A	na	na
ABo	A	A	A	na

Note that group O can give blood to any group and is referred to as the *universal donor*; group AB is the *universal recipient* because it can receive blood from donors of any group.

Rhesus factor

Another important blood group is the Rh factor. About 85 per cent of the population have Rhesus positive blood because their erythrocytes contain an antigen called the Rhesus factor. The remaining 15 per cent of the population have erythrocytes that lack the Rhesus antigen and their blood is called Rh-negative. If the Rh-positive blood is transfused into an Rh-negative patient the person responds by producing Rh-positive antibodies, that is the only effect. If, however, the Rh-negative recipient is given another dose of Rh-positive blood his Rhesus antibodies cause agglutination of the donor erythrocytes and can cause death.

In pregnancy, a Rh-negative mother can carry a Rh-positive child. If, at the end of pregnancy, minute portions of the foetal erythrocytes pass across the placenta and enter the maternal blood stream the maternal tissues respond by producing Rh antibodies which pass into the foetal circulation and destroy the erythrocytes, causing jaundice. The antibodies do not usually affect the first child seriously but if later children are Rh-positive there is a massive destruction of erythrocytes, a disease known as erythroblastosis foetalis or haemolytic disease of the new-born child. Treatment involves replacement of the baby's blood with Rh-negative blood. Prevention of the disease can be carried out by treating the mother with anti-Rh globulin which coats the foetal cells, so blocking the Rh factor.

SELF-ASSESSMENT QUESTIONS

1. Haemoglobin is a component of:
   (a) lymphocytes;
   (b) erythrocytes;
   (c) thrombocytes;
   (d) plasma;
   (e) monocytes.

2. Erythrocytes are made in the:
   (a) kidney;
   (b) spleen;
   (c) liver;
   (d) bone marrow;
   (e) lymph glands.

3. Erythrocytes are destroyed by the spleen at the end of their life which is about:
   (a) 300 days;
   (b) 50 days;
   (c) 12 days;
   (d) 120 days;
   (e) 22 days.

4. Which of the following fights infection in a septic wound?
   (a) erythrocyte;
   (b) thrombocyte;
   (c) plasma;
   (d) leucocyte;
   (e) lymph.

5. A person is called a universal recipient if he has blood group type:
   (a) A;
   (b) B;
   (c) AB;
   (d) O;
   (e) Rh-positive.

6. A person is called a universal donor if he has blood group type:
   (a) O;
   (b) AB;
   (c) A;
   (d) B;
   (e) Rh-negative.

7. Carbon dioxide is carried in the blood by:
   (a) plasma;
   (b) thrombocytes;
   (c) monocytes;
   (d) lymphocytes;
   (e) basophils.

THE BLOOD
ASSIGNMENTS

1. Find out how you would ascertain your own blood group and write up an account of the method.

2. Why is blood typing so important?

CHAPTER THREE

The Circulation of the Body Fluids

CHAPTER OBJECTIVES

After studying this chapter you should be able to:
* explain what is meant by the main body fluid compartments of the body;
* show how the fluid circulates between them;
* compare the physical properties and chemical composition of plasma, lymph and tissue fluid;
* describe the arrangement of the arteries and veins in the mammalian body;
* define the term "blood-pressure";
* state the normal blood-pressure between arteries and veins.

INTRODUCTION

This chapter enables the student to understand the role of the body fluids within the body; how they are produced and how they are specialised to perform particular functions, how they are interrelated and circulated.

The circulatory system consists of a circulatory fluid, the blood, which is pumped through a closed system of vessels by a pump, the heart. Thick-walled arteries take blood from the heart to the tissues where they subdivide into smaller and smaller vessels (arterioles) which eventually lead into the capillary network which provides a large surface area for the exchange of materials to and from the blood. The arterioles are fine tubes whose walls are composed of involuntary muscle fibres which run circularly around them. When these contract they constrict the calibre of the arteriole, and when they relax they dilate the arterioles and more blood flows into the capillaries. The greater the constriction the greater will be the resistance to the flow of the blood and the arterial pressure will rise. If the arterioles dilate more blood will pass into the capillaries and the arterial pressure will fall.

37

Capillaries have a smaller bore than arterioles and a larger surface area so there is more friction as the blood passes from the arterioles into the first part of the capillary bed. The friction lowers the blood-pressure still more. Blood from the capillaries drains into the wider venules which lead into the veins taking the blood back to the right atrium. This is the *systemic circulation*. Blood then passes into the right ventricle, into the pulmonary artery to the lungs where oxygenation occurs. The oxygenated blood is then taken back to the left atrium. This is the *pulmonary circulation*. The systemic and pulmonary circulations together constitute the *double circulation*.

THE HEART

The four-chambered heart (*see* Figs. 26 and 27) is a very efficient pump which possesses valves to ensure that blood is pumped in one direction only. The tricuspid valve separates the thin-walled right atrium from the thick-walled right

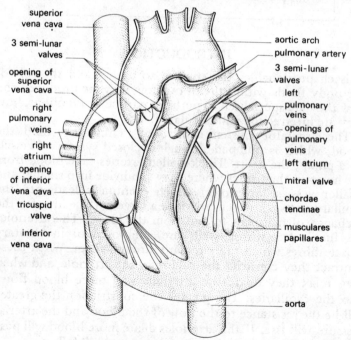

Fig. 26. *The internal structure of the heart.*

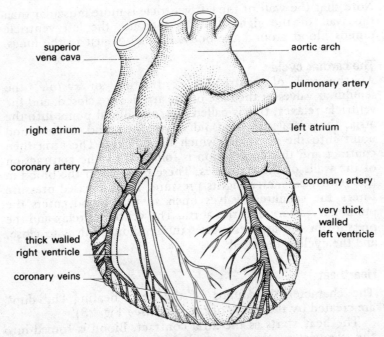

superior vena cava

aortic arch

pulmonary artery

right atrium

left atrium

coronary artery

coronary artery

very thick walled left ventricle

thick walled right ventricle

coronary veins

Fig. 27. *Ventral view of the dissected heart.*

ventricle; the bicuspid or mitral valve separates the thin-walled left atrium from the very thick-walled left ventricle. These valves only open into the ventricles and they are closed by the pressure of the blood in the ventricles. The tricuspid and bicuspid valves are attached to muscular pillars, called the musculares papillares, in the walls of the ventricles by tough cords called the chordae tendinae. These strong cords prevent the valves from being forced upwards into the atria by the great pressure of the blood when the ventricles contract. When this contraction occurs the pressure in the ventricles rises, the tricuspid and mitral valves close, the pressure increases and the semilunar valves which guard the entrance to the aorta and pulmonary artery, are forced to open so that blood rushes into the pulmonary artery and aorta. The ventricles relax again and the pressure in them falls to below that in the pulmonary artery and aorta. When this happens the valves at the base of the arteries close. Pressure falls further in the ventricles until it is below that of the atria.

Note that the wall of the left ventricle is more muscular than the wall of the right ventricle because the left ventricle pumps blood around the body, instead of just to the lungs.

The cardiac cycle

At the end of ventricular contraction or *systole,* the semilunar valves at the base of the arteries are closed, and the ventricle relaxes; this is called *diastole.* Blood pours into the atria, the mitral and tricuspid valves open, and so the blood pours into the ventricles which fill quickly. The atria then contract and this contraction is followed by the contraction of the walls of the ventricles. These press upon the blood in the ventricles, increasing its pressure. The increased pressure forces the semilunar valves open so that blood enters the pulmonary artery and the aorta. The ventricles relax and the pressure in the arteries causes the semilunar valves to close, and the cycle is repeated.

Heartbeat

The characteristic sounds of the heart beating "lub-dup" are created by the valves of the heart (*see* Fig. 28).

The beat starts as the atria contract. Blood is forced into the ventricles, the tricuspid and mitral valves contract producing the "lub" sound, and when the ventricles, contracting, force the blood past the semilunar valves, the valves close producing the "dup" sound. The average number of beats per minute is seventy-two in a resting person; this number varies with the age of the person and with the internal and external conditions, e.g. in illness, or with rise or fall in external temperature. Each beat pumps about 60 cm^3 of blood from the heart.

Initiation of heartbeat

Heartbeat is initiated in the sinoatrial node, a specialised muscular tissue, situated in the right atrium just above the entrance of the superior vena cava. Stimulation spreads out from this node to the atria which contract. The sino-atrial node in turn stimulates another node of specialised tissue situated at the junction of atria and ventricles; this is the atrio-ventricular node. Stimulation from this node spreads to the ventricles through Purkinje's fibres, specialised cardiac muscle fibres which run down the interventricular septum and spread out over the ventricular walls. Excitation of this tissue causes contraction of the ventricles.

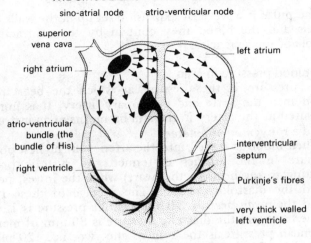

Fig. 28. *How the heartbeat is controlled.*

The heartbeat so initiated is controlled by the nervous system. Slowing of the heartbeat is brought about by efferent fibres of the vagus (Xth cranial) nerve which acts on the nodes, and acceleration of the heartbeat is brought about by cardiac sympathetic fibres which act directly on the ventricular muscle and also stimulate the sino-atrial node, and so the sino-atrial node acts as a pacemaker, ultimately controlling the rate of the beat of the heart.

The wall of the aorta and the walls of the right atrium both contain sensory cells which transmit impulses via the afferent fibres of the vagus; similarly, sensory cells in the carotid sinus (in the neck region) send impulses along the sinus nerve, all these afferent fibres being stimulated by increased blood-pressure. Their axons terminate in the cardiac control centre of the brain and so heartbeat is regulated to meet the needs and activities of the individual.

Other factors such as temperature (fever), pH (affected by carbonic acid concentration in the blood), hormones (such as adrenaline), can all affect heartbeat.

The pulse
Each beat of the heart causes a pulse in the large arteries which can be felt by placing the middle finger over a large artery, usually the radial artery in the wrist.

The pulse is a wave of expansion set up in the walls of the vessels and the blood they contain by the contraction or systole of the ventricles.

The blood-pressure in Man

Blood-pressure is the pressure at which the heart pumps blood into the aorta and pulmonary artery. It is normally measured at the artery in the arm by means of an instrument called a *sphygmomanometer.*

When blood is forced into the arteries the highest point of pressure in the arteries is termed the systolic pressure (relating to the beats of the heart) while the lowest point is called the diastolic pressure. The average of these two is called the *mean pressure.* If the systolic pressure is 120 mm of mercury and the diastolic pressure is 80 mm of mercury, the mean pressure is the sum of the two, i.e. 120 plus 80 divided by 2, or 100 mm of mercury. The pulse pressure is the difference between systolic and diastolic pressure; in this case 40 mm of mercury.

Factors affecting blood-pressure

(*a*) The pumping of the heart maintains the arterial pressure. If the heart discharges more blood into the arteries the arterial pressure rises. If the heart pumps out less blood the blood-pressure will fall.

(*b*) Arterial walls contain involuntary muscles and so they can constrict or dilate. Constriction increases resistance to the flow and causes the blood-pressure to rise in them. Dilation has the reverse effect.

(*c*) The elasticity of the arterial walls prevents the pressure falling too low between the beats of the heart.

(*d*) The viscosity of the blood will affect the flow of blood, the more viscous the blood the slower the flow and the greater the blood-pressure.

(*e*) The amount of blood in the arterial system affects the pressure. The more blood in the arteries the more their walls are stretched and the greater the blood-pressure. If blood is lost, e.g. in haemorrhage, the pressure drops.

Pressure in the capillaries

Pressure at the arteriolar ends of the capillaries is approximately 40 mm of mercury. This decreases to about 10 mm of mercury at the vein ends of the vessels. The capillaries have a fine bore and a large surface area which is the origin of the friction so that the speed of the flow of the blood is decreased. The result is a decrease in the blood-pressure.

Pressure in the veins

Blood flows from the capillaries into the veins. Capillaries and veins form a low pressure system as opposed to the high pressure system of the arteries and arterioles. In the venules and veins the surface area is smaller than in the capillary beds so there is less friction. The veins flow together to form larger veins and as the blood passes through the veins there is less resistance to flow and the pressure falls even further and it may fall to nearly zero as the blood reaches the right atrium whose expansion helps draw the blood into the heart. The velocity of the blood increases as the surface area decreases; the skeletal muscles by their normal contraction act as a "venous pump" for they compress the blood within the veins and this helps speed the flow of blood in the veins. This is the reason why patients are encouraged to use their muscles and not lie in bed for long periods during times of illness.

The veins contain valves inside them which prevent backflow of blood, particularly in the veins of the leg; the breaking down of these valves gives rise to a condition known as varicose veins.

Inspiration also affects blood-pressure in the veins. Inspiration is brought about by the contraction of the inter-costal muscles, which raises the ribs, and the flattening of the diaphragm. This combined action increases the size of the thoracic cavity. This tends to decrease the pressure in the thorax and expand the thin-walled venae cavae. Thus blood is "drawn into" the thorax. Relaxation of the ventricles during the cardiac cycle draws blood from the atrium and thus is then replaced by blood from the venae cavae.

HISTOLOGY OF THE BLOOD-VESSELS

Arteries

Arteries (*see* Fig. 29) have thick walls which enable them to withstand the pumping action of the heart. The internal lining is composed of squamous or pavement epithelium called endothelium. This is surrounded by an inner coat or tunica intima which is composed of elastic connective tissue.

The middle coat or tunica media is composed of unstriated muscle fibres, which are shorter than those found in the viscera, with long rod-shaped nuclei. The outer coat or tunica adventitia consists of areolar connective tissue and it contains many elastic fibres. This coat gives strength to the artery wall enabling it to resist undue expansion. The wall of the aorta differs from that of other arteries. It is lined by endothelium but the tunica intima is a thick layer of connective tissue rich in elastic fibres; it merges into the tunica media where a large amount of elastic tissue forms layers and networks which alternate with unstriated muscle.

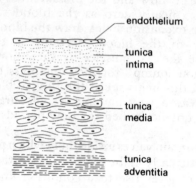

Fig. 29. *Section through artery wall.*

This large amount of connective tissue makes the aorta very strong and able to withstand the pressure of the blood as it is pumped from the heart. Outside this layer there is an outer coat, the tunica adventitia, containing fine elastic fibres.

Veins

Veins (*see* Fig. 30) are lined with endothelial tissue and have a tunica intima which contains far less elastic tissue than is found in arterial walls. The middle coat, the tunica media, is also less elastic and contains much less muscle tissue than is found in the walls of the arteries. The outer coat, the tunica adventitia is relatively better developed in veins than in arteries. This means that the walls of the veins are strong even though they are thinner than those of arteries.

Limb veins possess valves (*see* Fig. 31). The veins in the viscera, tributaries of the portal vein, veins in the cranium and vertebral column, the veins of bones and the umbilical vein do not possess valves.

The valves are pocket-like folds of the tunica intima strengthened by fibrous tissue. The tunica intima is thicker on the side subject to friction from the current of the blood flow and the endothelial cells are elongated on that side.

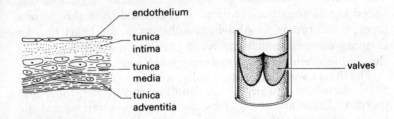

Fig. 30. *Section through wall of vein.*   Fig. 31. *Vein cut open showing valves.*

Capillaries

Capillaries (*see* Fig. 32) have walls composed of flattened endothelial cells only, which are continuous with those lining the arteries and veins. The diameter of capillaries varies from 8–20 μm which, at its smallest, is the same size as the erythrocytes and so friction is produced as the blood cells come into contact with the walls.

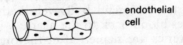

Fig. 32. *A capillary.*

BODY FLUIDS AND CIRCULATION

The flow of blood

The flow of blood is greatest in the arteries where the motive power is the pumping action of the heart. Blood flow is slower in the veins and slowest in the capillaries.

In all three types of vessels the flow is greatest in the centre and slowest near the walls, due to friction, and this is where leucocytes tend to be carried. If inflammation occurs in a tissue the leucocytes adhere to the walls and even pass through them into the surrounding connective tissue as migratory cells which can engulf and destroy the bacteria and other agents causing the inflammation. Thrombocytes also pass along nearest the walls.

Description of the circulatory system of man

The heart is contained in the thorax within a fibrous bag called the fibrous pericardium. This is lined by a thin serous layer, a continuation of which adheres to the heart muscle forming the pericardial cavity. It contains a small amount of fluid which reduces friction during heart beat.

The heart walls are continually working (i.e. beating) and they therefore require a continuous supply of food and oxygen. These are brought by the coronary arteries, and the waste products of the heart's metabolism, which include carbon dioxide, are carried away by the coronary veins.

The pumping heart receives the blood draining back from the body and lungs through thin-walled veins and it pumps blood to the tissues through thick-walled arteries, so, in general, arteries contain oxygenated blood whereas veins carry deoxygenated blood. However, exceptions to this are the pulmonary arteries, which carry deoxygenated blood, and the pulmonary veins, which carry oxygenated blood. These veins bring oxygenated blood back to the left side of the heart and from here it is pumped through arteries to all parts of the body. The main trunk of the arterial system is the very thick walled aorta. The main body vein is the inferior vena cava which carries blood back to the heart.

Veins and arteries are named after the organs they serve, thus the renal artery carries oxygenated blood, laden with nitrogenous waste, to the kidney and the thin-walled renal vein carries the deoxygenated blood, now purified, to the inferior vena cava.

This knowledge of the system of naming, together with a diagram (see Fig. 33 and 34), enables the description of the circulation in man to be quite simple but there is a notable exception to all this: the *liver*. This important organ has *three* blood vessels serving it (see Fig. 35); the hepatic artery carrying oxygenated blood to its tissues; the hepatic vein carrying deoxygenated blood from the liver to the vena cava; and the *hepatic portal vein* carrying blood laden with glucose and amino acids from the capillaries of the small intestine to the capillaries of the liver tissues (see Chapter 7).

To summarise: arteries carry blood from the heart to the tissues; veins carry blood from the tissues to the heart, communication between the two occurs in the capillary bed:

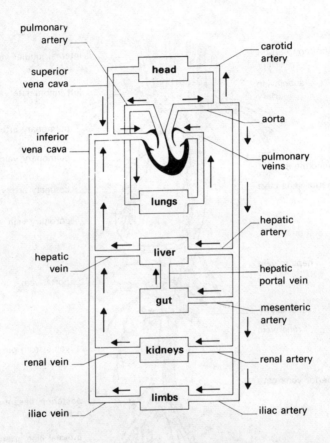

Fig. 33. *The blood vascular system.*

heart ———→ arteries ———→ arterioles ———→ capillary ➔
➔ bed ———→ venules ———→ veins ———→ heart.

The hepatic portal system carries blood from one set of capillaries to another set of capillaries:

capillaries of intestinal wall ————→ hepatic portal vein ➔
————————→ capillaries of the liver.

Remember that the pulmonary vein is the only vein which carries oxygenated blood and the pulmonary artery is the only artery that carries deoxygenated blood.

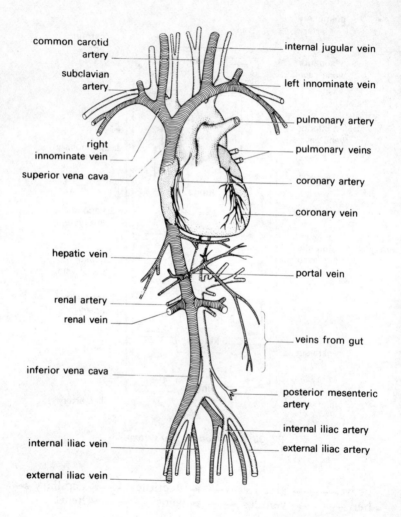

Fig. 34. *Vascular system of man.*

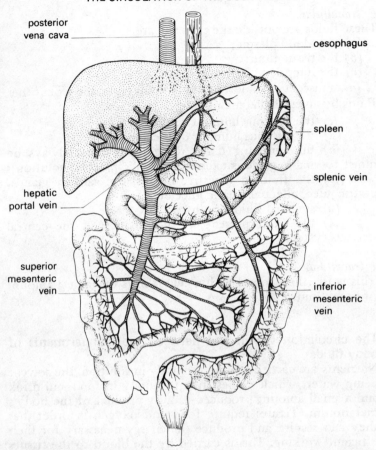

posterior
vena cava

oesophagus

spleen

splenic vein

hepatic
portal vein

superior
mesenteric
vein

inferior
mesenteric
vein

Fig. 35. *The hepatic portal system.*

The compartments of the body fluids

Blood is not the only body fluid, it is one of the three body
fluids that provide the internal environment for the cells of
the body and make up approximately 60 per cent of the
body weight. Nutrients are supplied to the tissues by the
body fluids and excretory products are removed by them. It
follows then that the composition of the body fluid must be
regulated in order to maintain life. The two main compartments of body fluids are as follows.

Extracellular

These fluids are outside the cells and are:

   (*a*) the blood plasma;

   (*b*) the tissue fluid;

   (*c*) the lymph;

   (*d*) transcellular fluids, i.e. body fluids separated from other fluids by an epithelial membrane:

     (*i*) the aqueous humour of the eye;

     (*ii*) the vitreous humour of the eye;

     (*iii*) fluid in lung tissue (lung tissue must always be moist to enable gaseous exchange to take place in solution);

     (*iv*) the fluid in digestive organs, e.g. salivary glands, gastric juice, bile, pancreatic juice etc.;

     (*v*) synovial fluid in joints;

     (*vi*) the fluid present in the cavities of the central nervous system.

Intracellular

This is the fluid found within cells; this is the largest amount of body fluid.

The circulation within and between the compartments of body fluids

Nutrients are carried around the body in solution, the solvent being water, which comes into the body in food and drink and a small amount produced as a by-product of the body's metabolism. Tissues require food and oxygen in order that they can respire and produce the energy necessary for their efficient working. This is carried by the blood to the tissues from which it carries carbon dioxide and other waste substances away. The compartments of body fluids are surrounded by tissues which are in communication with the circulatory system via the capillaries. Water, food, oxygen and salts pass out of the blood capillaries into the tissue fluid which bathes the tissue cells. Tissue fluid drains into the lymphatic vessels and can pass back into the blood-stream via the thoracic duct. The water balance of the body must be maintained; water enters the plasma when it is absorbed from the alimentary canal (from the food) or from the tissues (produced from tissue repiration) and it is lost from the plasma through the skin, lungs, kidneys and rectum (in the faeces). Since blood plasma, tissue fluid, lymph and tissues are all in communication with each other, all the body fluid compartments are linked.

Plasma, lymph and tissue fluid are all nearly identical in the concentration and composition of electrolytes which are mostly sodium and chloride. Intracellular fluid contains more protein than extracellular fluids and its primary electrolytes are potassium and phosphate. Plasma has the highest protein content of all the tissue fluids. As would be expected the composition of transcellular fluid varies greatly because each fluid is produced for a specific function, e.g. the gastric juice contains the enzymes for digestion in the stomach, the synovial fluid is produced to lubricate the joint. Nevertheless, all these fluids receive nutrients from the plasma of the blood and waste products from the tissues they serve which ultimately return to the plasma either direct or via the tissue fluid lymphatic system route.

Oedema

Fluid can accumulate in the connective tissue compartment of the body fluids causing swelling, or oedema, which can occur when filtration from the blood vessels exceeds the osmotic return or when there is retention of salt and water or the lymphatic vessels are blocked. Another cause can be failure of the heart to pump effectively. This means that the pressure in the arteries is lower than normal and the blood flow is slower. When blood reaches the capillary bed it is slowed down again and the pressure becomes very low there. Tissue fluid seeps into the tissues and it accumulates there, aided by the pull of gravity. The patient's tissues cannot get sufficient food and oxygen and this upsets the metabolism still further. Fluid accumulates in the tissues of the lungs and oxygenation is impaired. The patient has difficulty in breathing, cannot move about and therefore the voluntary muscle action cannot aid the flow of blood in the veins and lymph in this impedes heart beat still further; this happens in chronic heart failure.

Plasma

Plasma is a straw-coloured fluid which forms a specialised matrix of blood in which erythrocytes (or red corpuscles), leucocytes (or white corpuscles) and blood platelets (or thrombocytes) float. Plasma is composed of 90 per cent water, containing the proteins serum albumin, globulin and fibrinogen in colloidal solution; these buffer the blood against violent changes in pH. Fibrinogen and prothrombin are concerned with clotting of blood. Plasma transports materials around the body and everything that is carried around the body can be found in the plasma at some stage or other, e.g. hormones are present on their way from the endocrine organs that produce them to the glands or structures they stimulate; the gases oxygen, carbon dioxide (dissolved as bicarbonate ions) and nitrogen are there in the plasma together with nitrogenous waste, urea; the products of digestion, glucose, amino acids, minute fat globules are there as well as antibodies and antitoxins. The salts present include those of iron, calcium, potassium and sodium as chlorides, phosphates and carbonates.

Tissue fluid

Blood is pumped through the arteries by the action of the heart, into the arterioles and finally into the capillaries. The capillary walls offer resistance to the flow of blood and this forces materials out from the blood through the capillary wall under hydrostatic pressure.

The capillary walls are permeable to water, glucose, salts, amino acids, vitamins and hormones, and these materials are known collectively as tissue, or interstitial fluid. Blood proteins, red corpuscles and thrombocytes do not pass out of the capillary. Tissue fluid bathes the tissues and acts as an efficient intermediary between the blood and the tissues. Some of the tissue fluid drains back into the capillaries but the rest drains into a network of tubular vessels whose walls are only one endothelial cell thick, the lymphatic vessels.

Lymph

When the tissue fluid drains into the lymphatics it becomes known as lymph. Lymph contains about the same concentration of diffusible substances as plasma but its protein concentration is lower because the filtration of proteins is slower than that of other substances. Lymph vessels in the small intestine (called lacteals) may contain lymph with a high concentration of absorbed fatty acids and glycerol absorbed from a digested meal (*see* p. 99). Lymph is drained into tiny lymphatics which drain into large vessels which finally communicate with the blood via the thoracic duct. The main thoracic duct empties into the left innominate vein in the region of the neck. This duct drains the abdomen and thorax whilst a smaller duct drains the head and the right arm via the right innominate vein.

Lymph moves slowly and steadily around the body; this movement is due mainly to the constant formation of tissue fluid from the capillaries but the action of the muscles and the movements of respiration also help to move the lymph along, and like veins, the lymphatic vessels possess valves to prevent backflow. If the muscular tissues relax around the lymphatics they expand and draw fluid into the lymphatic vessels and contraction of muscles compresses the lymphatics and moves the fluid on. The movement of muscles occuring during sleep ensures that the lymph continues to flow. Lymph returns excess fluid and filtered protein to the blood stream; it is also concerned with defense and leucocyte production. Most of the medium-sized lymphatics possess rounded structures known as lymph nodes at intervals along their length. Each of these nodes possesses an incoming afferent lymphatic and an outgoing efferent lymphatic. Fibrous tissue surrounds the node and further connective tissue makes up its body. The lymph passes through these lymph nodes via lymph sinuses which contain two types of cells.

(*a*) *Reticular cells* give rise to lymphocytes and plasma cells; the lymphocytes are shed into the efferent lymph flow and the plasma cells, areas of antibody production, remain in the nodes or pass out to other areas.

(*b*) *Macrophages* engulf bacteria and cancer cells. These large phagocytic cells may be fixed in the lymph node or may wander into the tissues; in both situations they ingest bacteria.

Lymphoid tissue is also found in the tonsils, adenoids, spleen and in the Peyer's patches (these are roundish patches of aggregated lymph nodules on the intestinal walls). Lymphoid tissues can be regarded as safety barriers which prevent infective matter entering the blood from the tissues. This becomes obvious in extreme cases of infection, e.g. a septic finger soon causes swelling to occur in the lymph glands present in the armpits as the lymphocytes "gobble up" the bacteria. In the case of cancer of the breast, its removal may involve the removal of the lymph nodes and network from that region of the body and sometimes this results in the arms and hands becoming swollen because there is ineffective drainage of the lymph back from the tissues.

Table II compares the sites, functions, movement and composition of body fluids while Table III summarises how carbon dioxide is carried in the blood and lymph.

TABLE II. COMPARISON OF PLASMA, TISSUE FLUID AND LYMPH

	Plasma	Tissue fluid	Lymph
Where found.	Fluid matrix of blood. Passes through heart, arteries, arterioles, capillaries, venules and veins.	Bathes tissues.	Contained within the lymphatic vessels.
Function.	Transports substances around the body.	Acts as a "go-between", linking capillaries and tissues.	Drains excess fluid from the tissues together with proteins. Defensive — lymph nodes produce lymphocytes and macrophages.

	Plasma	Tissue fluid	Lymph
Movement caused by:	beat of the heart.	hydrostatic pressure of capillaries and drainage by lymphatics.	production of tissue fluid, movement of muscles and other tissues.
Composition.	All the constituents of blood.	Lower protein content than plasma. No erythrocytes and no thrombocytes.	Similar to tissue fluid but contains a higher proportion of leucocytes than plasma.

TABLE III. HOW CARBON DIOXIDE IS CARRIED
IN BLOOD AND LYMPH

Blood	Lymph
5 per cent in simple solution; the rest carried chemically combined in plasma and in erythrocytes.	Carried in simple solution.

SELF-ASSESSMENT QUESTIONS

1. Lymph is found in:
   (a) the gastric juice;
   (b) the eye;
   (c) veins;
   (d) blood clots;
   (e) lymphatic vessels.

2. Lymph enters the circulatory system through:
   (a) the pulmonary artery;
   (b) the pulmonary vein;
   (c) the superior vena cava;

(d) the thoracic duct;
(e) the portal vein.

3. Which of the following organs has two veins and one artery associated with it?
    (a) stomach;
    (b) liver;
    (c) spleen;
    (d) ureter;
    (e) kidney.

4. Which two of the following vessels possess valves?
(i) arteries; *(ii)* veins; *(iii)* lymphatics; *(iv)* capillaries.
    (a) *(i)*, *(ii)*;
    (b) *(ii)*, *(iii)*;
    (c) *(iii)*, *(iv)*;
    (d) *(ii)*, *(iv)*;
    (e) *(i)*, *(iv)*;

5. The blood vessel which passes blood into the right atrium is called:
    (a) the aorta;
    (b) the pulmonary vein;
    (c) the vena cava;
    (d) the hepatic portal vein;
    (e) the hepatic vein.

6. The blood vessel which takes blood from the left ventricle is:
    (a) the pulmonary artery;
    (b) the pulmonary vein;
    (c) the vena cava;
    (d) the aorta;
    (e) the hepatic portal vein.

7. The blood vessels containing the most glucose come from the:
    (a) liver;
    (b) kidneys;
    (c) intestines;
    (d) legs;
    (e) arms.

8. Which of the following blood vessels carries oxygenated blood?

- (a) superior vena cava;
- (b) inferior vena cava;
- (c) pulmonary artery;
- (d) pulmonary vein;
- (e) renal vein.

ASSIGNMENTS

1. Take your pulse for one minute by using the third finger of your right hand and compress the radial artery of the other hand against the bone. Carry out some vigorous exercise, then take your pulse again. What is the difference? Make a graph showing the time taken for your pulse rate to return to normal.

2. Draw a large labelled diagram of a human heart. Colour blue those structures containing deoxygenated blood. Prepare a table showing the function of *all* the parts.

CHAPTER FOUR

Respiration

CHAPTER OBJECTIVES

After studying this chapter you should be able to:
* describe the mechanism of quiet respiration;
* describe the modifications that can increase alveolar ventilation;
* identify histological sections of normal lung tissue;
* discuss the structure of the lungs and associated air passages in relation to their function;
* define the terms tidal volume, minute volume and vital capacity;
* state the normal values for tidal volume, minute volume and vital capacity;
* describe a method for measuring the rate and depth of respiration in man.

INTRODUCTION

All body cells require energy, which is obtained by the oxidation of food material in the mitochondria of the cells. Therefore, all the body cells require oxygen. This gas is readily available in the air and it is taken into the body during breathing. The oxidation of food not only produces the required energy but also results in the formation of large amounts of carbon dioxide. Therefore oxygen must be taken into the body and carbon dioxide must be expelled. This exchange of gases takes place by diffusion over the large, moist surface area of the lungs, a process of diffusion which must be rapid in order that the blood may be oxygenated as it passes through the lungs. In order that this may be so, the concentration gradient of the gases at the lung surface is kept high by the muscular action of the diaphragm and intercostal muscles. Blood brings carbon dioxide from the tissues to the alveoli (*see* Fig. 36).

The lungs are passive structures; during inspiration air is drawn into them by the intercostal muscles contracting, and the raising of the ribs, so increasing the capacity of the thoracic cavity; this process is accompanied by the contraction of the diaphragm. The reverse process is

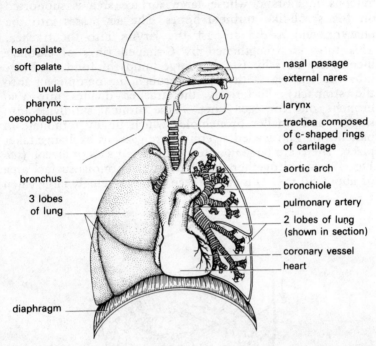

hard palate
soft palate
uvula
pharynx
oesophagus
nasal passage
external nares
larynx
trachea composed
of c-shaped rings
of cartilage
bronchus
3 lobes
of lung
aortic arch
bronchiole
pulmonary artery
2 lobes of lung
(shown in section)
coronary vessel
heart
diaphragm

Fig. 36. *The respiratory system, ventral view.*

expiration: the intercostal muscles relax and lower the ribs, the diaphragm domes and carbon dioxide-laden air is pushed out of the lungs.

Internal and external respiration

The process whereby oxygen is taken into the body through the alveoli is called external respiration, but this is only one part of respiration. The second part occurs when the oxygen which has been brought to the cells in the erythrocytes is used to oxidise the food and produce energy. This is internal or tissue respiration.

THE ANATOMY OF THE RESPIRATORY SYSTEM

Air is breathed in through the external nares, or nostrils. These holes lead into the nasal cavities which are lined by moist mucous membrane and hairs which filter out dirt. The air is also warmed and moistened as it passes over the mucous membrane whose large surface area is supported on the scroll-like turbinal bones. The air passes into the pharynx and is drawn past the larynx into the trachea. This tube is strengthened by C-shaped rings of cartilage incomplete dorsally (so that large lumps of food can bulge into the trachea as they pass through the oesophagus into the stomach). Posteriorly the trachea divides into two bronchi, one for each lung. These bronchi subdivide into smaller and smaller bronchioles which finally terminate in bunches of thin walled alveoli where gaseous exchange takes place. There are large numbers of these air sacs or alveoli (*see* Fig. 37) and altogether they have the enormous surface area of about 50 m². This can be seen to be extremely large when

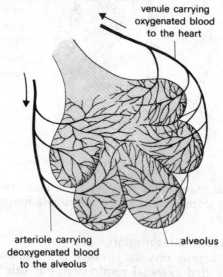

venule carrying
oxygenated blood
to the heart

arteriole carrying
deoxygenated blood
to the alveolus

alveolus

Fig. 37. *The alveoli and their blood supply.*

compared with the average surface area of an adult's body which is 2 m². These very thin, moist alveoli have many blood capillaries which form networks near their surfaces; this ensures that fast diffusion of gases can take place, i.e. oxygen from the air in the alveolus into the blood and carbon dioxide from the blood into the air in the alveolus.

The alveoli have some elasticity enabling them to cope with the changes in volume that occur during breathing. The lungs are not 100 per cent efficient and not all the air breathed in is breathed out; some remains in parts of the lung where no gaseous exchange takes place. In fact the alveoli may receive only 60 per cent of the total inhaled air.

Tidal volume

The amount of air taken in during breathing is known as the tidal volume. Man at rest takes in about 500 cm^3 of air per breath, thus his tidal volume is 500 cm^3.

Minute volume

This is the total amount of air entering the lungs per minute. At rest man takes approximately sixteen breaths per minute; if each contains 500 cm^3 of air, the minute volume will be 16 × 500 cm^3, i.e. 8000 cm^3, that is, the lungs are ventilated by 8 litres of air per minute.

The vital capacity

The vital capacity is the largest volume of air that can be expired after the deepest possible inspiration; in a fit man it can be as much as 4 litres.

THE LUNGS AND GASEOUS EXCHANGE

The lungs have a natural elasticity due to the elasticity of the alveoli; because of this the lungs would tend to contract at expiration if it were not for the serous pleural membranes which cover the lungs and are reverted back over the inside of the thoracic cavity forming a closed system. These moist membranes prevent friction during breathing movements. The lung's elasticity facilitates respiration in a great variety of conditions, e.g. gentle breathing when the body is at rest, panting after vigorous exercise.

Oxygen dissolves in the moist film present on the walls of the alveoli and diffuses through the walls into the capillaries, where it combines with the haemoglobin in the erythrocytes forming oxyhaemoglobin. The concentration of oxygen in the blood of the lung capillaries is less than that in the alveoli and so oxygen will always diffuse from the alveoli into the blood. Carbon dioxide is brought to the alveoli as bicarbonate ions in the blood plasma. In the lung tissue enzymes produce carbon dioxide from the bicarbonate ions and this diffuses along a concentration gradient from the blood into the alveoli from which it is expired into the atmosphere.

Histology of the lungs

The larynx, or voice-box, is found at the posterior end of the pharynx, ventral to the oesophagus; it leads into the trachea posteriorly.

The trachea, or windpipe, is a tube containing fibrous and muscular tissue which is kept open by the C-shaped rings of cartilage; the tube is lined by ciliated epithelium, containing goblet cells which secrete mucus, to which dirt particles adhere and they are then wafted up to the pharynx by the cilia where they are either spat out or swallowed.

Posteriorly, the trachea divides into the two bronchi which subdivide into smaller and smaller bronchioles so forming a "bronchial tree". Bronchi and bronchioles have similar structures to that of the trachea and they too are supported by cartilage.

In the finest bronchioles the ciliated epithelium gives way to cubical epithelium and these bronchioles lead into the alveolar passages which communicate with blind passages or

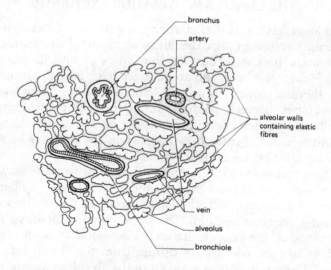

Fig. 38. *Section through lung tissue.*

infundibula from which the round alveoli open off; they have the appearance of minute bunches of grapes (*see* Fig. 37).

The walls of the alveoli contain groups of cubical epithelial cells together with very thin squamous epithelium which separate the blood in the blood capillaries from the inspired air in the air sacs. There is also some delicate connective tissue (elastic fibres) in the walls of the alveoli (*see* Fig. 38). Phagocytes which engulf bacteria and dirt particles, are present along the alveolar walls which have an extremely good blood supply.

Control of respiration
Respiration is governed by two means.

Nervous control
We can all modify our respiratory movements: we can hold our breath, breathe quickly, breathe slowly, breathe deeply, breathe lightly, sing, talk, etc. However, when we are asleep and while we are awake breathing carries on unconsciously and automatically. The impulses which control these movements come from the respiratory centre in the medulla oblongata of the brain. The medulla oblongata is the most posterior portion of the brain and is continuous with the spinal cord. Impulses from nerve cells in the medulla oblongata are sent out in a continuous stream to the muscles concerned with inspiration and expiration so that the thoracic cavity is constantly and rhythmically enlarged and reduced in volume. Damage to the respiratory centre by wounding, pressure, anaesthetics or drugs such as morphine prevents the impulses being sent to the respiratory muscles and can result in death.

The respiratory centre receives impulses from various parts of the body through various afferent nerves. This means that inspiration and expiration can be under reflex control and this is a survival factor, for example, if you smell something offensive, you suddenly find yourself holding your breath. Stimulation of sensory nerves within the body can affect the respiratory centre, e.g. taking a cold shower, having a hot bath.

If an irritating substance such as pepper comes into contact with the nasal mucosa it brings about the reflex act of sneezing, a deep inspiration which is followed by a violent expiration. If an irritant stimulates the sensory nerves in the larynx or trachea, e.g. a foreign body such as a piece of meat, cough-ing occurs to clear the air passage. This is a reflex action which occurs through the respiratory centre. There is first a quick inspiration followed by forcible expiration with the opening to the larynx closed; thus high pressure is built up within the lungs and with the sudden opening of the larynx the compressed air rushes out with the offending particle.

Crying, laughing and yawning are all types of respiration modified by nervous impulses from the brain to the respira-tory centre. Grief, excitement and fear can also affect the respiratory centre in a similar way.

Chemical control of the respiratory centre

Tissue respiration involves the production of the waste products carbon dioxide and water. The carbon dioxide enters the blood and passes through the respiratory centre which it stimulates constantly to produce a continual flow of impulses to the intercostal muscles and the diaphragm. This affects respiration and is part of the control of the internal environ-ment called *homeostasis*.

An increase in the carbon dioxide content of the blood increases the rate and depth of breathing, a decrease in the carbon dioxide content slows the rate of respiration and reduces the depth of breathing. The carbon dioxide content of the blood in a healthy person is kept fairly constant. If we try to hold out breath for a long time we find that we are forced to inspire; this is because carbon dioxide accumulates in the blood and stimulates the respiratory centre which makes us inspire.

We all know that we pant after running. This is because our bodies start the run from the state where we are carrying out normal quiet breathing. When we run, food in the muscles is oxidised to produce energy and increased carbon dioxide in the blood which affects the respiratory centre. However, before this affects the respiratory rate, the muscle respiration uses all the available oxygen and so has to switch to anaerobic respiration and so produce lactic acid. This lactic acid is later taken to the liver and converted back into glycogen but this requires oxygen so, in order to repay this *oxygen debt*, we pant and continue to do so after exercise has ceased, and thus inhale extra oxygen.

Anoxia or hypoxia

Anoxia means "no oxygen" and is a term referring to the condition where tissues receive an inadequate oxygen supply. Hypoxia means "low oxygen" and is a more accurate term than anoxia for if there is literally no oxygen, the person will die. Hypoxia can occur in many conditions.

(a) The blood may not be sufficiently oxygenated in the lung which may be the result of disease, e.g. lung cancer or tuberculosis. It may, however, be caused by a low percentage of oxygen in the air. This happens at high altitudes and may result in mountain sickness.

(b) Carbon monoxide is taken up by haemoglobin which forms the stable compound, carboxyhaemoglobin. If a person breathes in carbon monoxide, carboxyhaemoglobin is produced and less haemoglobin is available to carry oxygen; the result is hypoxia which ends in death. (Tobacco smoke contains carbon monoxide.)

(c) Anaemia can result in hypoxia because there is insufficient haemoglobin available to carry the oxygen.

(d) In cases of heart disease the capillary circulation is slow and this results in hypoxia.

(e) Poisoning of the tissues by cyanide can impair the uptake of oxygen and result in hypoxia.

Hypoxia is a dangerous condition. All tissues require oxygen in order to oxidise food materials and produce energy for their metabolic activities. The brain and nervous tissues are most quickly and easily affected. The effect is impairment of vision together with a lack of judgment. The lack of oxygen in the tissues finally shows itself when the person becomes unconscious and dies.

Comparison between inspired and expired air

Air that is taken into the body contains about:

(a) 21 per cent oxygen by volume;
(b) 0.04 per cent carbon dioxide by volume;
(c) 79 per cent nitrogen by volume.

Expired air contains about:

(a) 16 per cent oxygen;
(b) 5 per cent carbon dioxide;
(c) 79 per cent nitrogen.

TISSUE RESPIRATION

Oxygen absorbed from the lung is carried in the erythrocytes and is combined with haemoglobin, as oxyhaemoglobin. In the mitochondria of the cells making up the tissues of the body, glucose, which has been carried to the cells by the plasma, is oxidised to form carbon dioxide and water:

$$C_6H_{12}O_6 + 6O_2 \longrightarrow 6CO_2 + 6H_2O + energy.$$

The whole reaction, however, is not as straightforward as the equation appears. It occurs in a series of steps.

First the glucose is oxidised to form pyruvic acid. This enters a complicated chemical cycle called Kreb's cycle. The pyruvic acid is further oxidised to form carbon dioxide and water; this oxidation process is carried out by dehydrogenase enzymes which remove hydrogen. The pyruvic acid breakdown and Kreb's cycle both require oxygen and are therefore said to be aerobic; and furthermore they both produce energy.

The energy released is used to convert adenosine diphosphate (ADP) into the more complicated energy-rich adenosine triphosphate (ATP). The adenosine triphosphate can readily give up its energy to produce adenosine diphosphate once more and the energy so released can be used to provide the heat energy to keep the body warm (another aspect of homeostasis) or to provide energy for metabolic reactions within the cell. ADP can be converted into more ATP when more energy is released by oxidation in tissue respiration.

MEASUREMENT OF THE RATE AND DEPTH OF BREATHING

The whole pattern of respiration can be studied by means of a recording spirometer (*see* Figs. 39 and 40). This consists of a mouthpiece connected to two wide-bore tubes, one of which carries the inspiratory air, the other carries the expiratory air. The spirometer is a cylindrical vessel with two walls, one inside the other. Between the two walls there is a spirometer bell which is counterbalanced by a pulley and weight. Water forms the seal between the outer and inner walls. Air is let into the bell until the pointer comes to zero reading on the graduated scale on the rotating drum, the recording pen is fitted in position and the spirometer is ready for use.

A clip is placed over the subject's nose and he is asked to breathe quietly through the mouthpiece. As this happens, the carbon dioxide is absorbed by the soda-lime and the bell moves up and down with each respiration. The pen traces the pattern of quiet respiration on the recording drum. After calibration for time and volume have been carried out the oxygen consumption of the subject can be calculated. This can be utilised to ascertain the person's metabolic rate. It can also be used to measure the *tidal volume* which is simply calculated by multiplying the number of breaths per minute by the volume taken in during a breath.

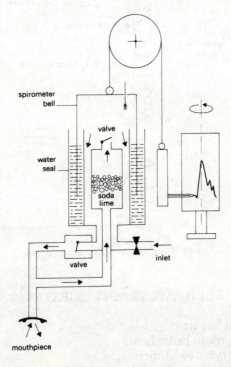

Fig. 39. *The spirometer.*

If a very deep breath is taken, the inspiratory reserve volume can be measured by the spirometer and if at the end of a normal expiration the subject is asked to breathe out as hard as possible the expiratory reserve volume can be measured. Inspiratory reserve volume plus tidal volume plus expiratory reserve volume together make up the vital capacity. However, even after the expiratory reserve volume has been forced out, there is still some *residual air* left in the lungs.

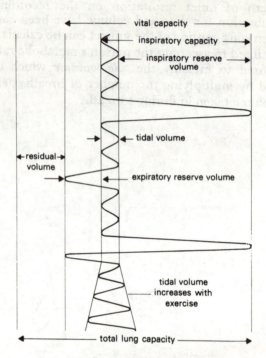

Fig. 40. *The different lung volumes in man.*

SELF-ASSESSMENT QUESTIONS

1. The trachea links:
   (a) larynx to bronchus;
   (b) pharynx to bronchus;
   (c) pharynx to stomach;
   (d) mouth to oesophagus;
   (e) bronchus to bronchioles.
2. The breathing rate is regulated by the brain:
   (a) entirely by voluntary control;
   (b) according to the pulse rate;

(c) according to the blood pressure in the arteries;

(d) according to the level of carbon dioxide in the blood;

(e) according to the level of the oxygen in the blood.

3. If you run a race you continue to breathe hard for several minutes afterwards because:

(a) it takes time for your pulse to slow down;

(b) you are tired;

(c) you are paying off the oxygen debt;

(d) it takes time for your respiration rate to slow down;

(e) you need to cool down.

4. In tissue respiration oxygen reacts with glucose to produce:

(a) energy, carbon dioxide and lactic acid;

(b) energy, carbon dioxide and water;

(c) water and carbon dioxide;

(d) carbon dioxide and energy;

(e) water, carbon dioxide and energy.

5. Another name for air sacs is:

(a) intercostals;

(b) alveoli;

(c) infundibula;

(d) bronchioles;

(e) bronchi.

6. Which of the following allows the lungs to move freely and without friction between the inner walls of the thorax?

(a) moisture in inhaled air;

(b) pulmonary blood supply;

(c) mucus in the bronchioles;

(d) fluid within the pericardium;

(e) fluid within the pleura.

7. The vital capacity is equal to:

(a) the tidal volume;

(b) the tidal volume plus the expiratory reserve volume;

(c) the tidal volume plus the inspiratory reserve capacity;

(d) inspiratory reserve capacity plus the expiratory reserve volume;

(e) the tidal volume plus the expiratory reserve volume plus the inspiratory reserve capacity.

ASSIGNMENTS

1. Count the number of breaths you take per minute. Do some violent exercise. Count the number of breaths per minute until the respiration rate returns to normal. Draw a graph of your result.

2. Write a short account of what has happened, explaining why it takes some time for the respiration rate to return to normal.

CHAPTER FIVE

Diet

CHAPTER OBJECTIVES

After studying this chapter you should be able to:
* state what is meant by a balanced diet:
* explain the need for a balanced diet both in terms of calorific values and chemical constituents;
* state what is meant by the term "micronutrient";
* state the source and explain the importance of minerals in the diet;
* state the source and explain the importance of vitamins in the diet;
* state the causes of the most common nutritional deficiency diseases.

INTRODUCTION

Food is required to provide the body with energy and the materials required for the various life processes such as growth, repair of cells, reproduction, providing heat, etc.

Nutrition involves the study of the intake of this food and its role in the processes of the growth, maintenance and repair of the body. Nutrients include carbohydrates, fats, proteins, vitamins, mineral salts and water; carbohydrates supply the body with energy and also may be laid down as fat; fats, too, may be deposited in adipose tissue for future use or may be used as a source of energy for present body needs; proteins provide the amino acids required for the growth and repair of tissues and vitamins are required in small amounts and are used to regulate body processes, as are mineral substances. Water is required as a solvent in such quantities as to make up two-thirds of the body weight.

A *balanced diet* contains carbohydrates, fats, proteins, vitamins, mineral salts and water in the correct proportions together with roughage which provides the "bulk" for efficient peristaltic action. Roughage is food with a high content of undigestable material consisting principally of cellulose.

FOOD AND ENERGY

Food contains stored energy. Green plants are called primary producers because they are able to utilise the light energy from the sun to build up the simple low energy molecules of water and carbon dioxide into high energy carbohydrate molecules. This process is called *photosynthesis* and is dependent on the green pigment chlorophyll, which is contained in special organelles in the leaves, called chloroplasts. The valuable by-product of this process is oxygen which replenishes the atmosphere and is used in aerobic respiration.

Carbohydrates produced in green plants by photosynthesis can be elaborated into other food molecules, e.g. fats, oils, vitamins, proteins, etc. This type of nutrition is called *holophytic* or *autotrophic*.

Animals, however, have *holozoic* nutrition; they cannot make high-energy complex molecules for themselves from very simple ones but have to take in food molecules elaborated by plants. Animals which feed entirely on plants are called primary consumers or herbivores, e.g. cow, sheep, rabbit, while animals which feed on other animals are called secondary consumers or carnivores. Man is included in a third group of animals and is classified as an omnivore, obtaining high energy food molecules from both plant and animal sources.

Carbohydrates

Carbohydrates are the chief source of energy in man's diet and they consist of carbon, hydrogen and oxygen atoms. The carbohydrates can be differentiated into three çategories according to the size of their molecules and reaction to certain reagents.

Monosaccharides

These are simple sugars which are soluble in water. The commonest are glucose and fructose which are to be found in fruits. Glucose is the final product of carbohydrate digestion in the gut of man; it is soluble and is carried around the body and is used by the mitochondria to supply energy by oxidation, i.e. in tissue respiration. Large amounts of glucose in solution could have deleterious osmotic effects in the cells of the body as it travels around the blood and so its concentration is maintained at a constant level by the action of the hormone, insulin. Insulin ensures that excess soluble glucose is converted into glycogen in the liver.

Monosaccharides are called reducing sugars because they reduce alkaline copper (II) salts, such as Fehling's solution to the orange precipitate of the copper (I) salt.

Disaccharides

These are more complex, consisting of two monosaccharide molecules chemically linked together. This linkage means that disaccharide molecules do not usually give a positive (orange precipitate) result with the Fehling's test. If the disaccharide is treated drastically by boiling with hydrochloric acid, hydrolysis occurs and the molecule is split into its two components which will then give a positive test for a reducing sugar. This hydrolysis occurs in the gut during digestion far more gently by means of enzyme action. *Sucrose*, or cane sugar, is the most familiar disaccharide which we eat; found in sugar cane and sugar beet in solution in the cell sap, it is purified and crystallised to produce the sugar we add to our beverages.

Polysaccharides

These are large complex molecules consisting of up to many thousands of simpler carbohydrate units joined together. Their complexity means that they are insoluble and therefore may be laid down as food storage material. In plants this is often in the form of starch laid down in specialised cell organelles called *leucoplasts,* which may be massed in special storage organs, e.g. potato tubers and wheat grains. Starch gives a characteristic blue-black coloration with iodine in potassium iodide solution. In man and most other animals, animal starch or glycogen is stored principally in the liver. Cellulose is another polysaccharide which is of some importance in man's diet; it forms the cell walls of plant cells and is therefore a large constituent of all fruits and vegetables. However, man cannot digest cellulose because he does not produce the necessary enzymes. It therefore passes through unchanged, providing bulk to the diet which stimulates the circular and longitudinal muscles in the gut walls to carry out effective peristalsis. This ensures that defaecation occurs regularly and often, so preventing bacteria building up toxic substances which could have deleterious effects on the cells of the gut walls.

Fats

Fats, or lipids, are very high energy compounds which also consist of carbon, hydrogen and oxygen atoms although they are present in different proportions from those found in carbohydrates. Fats are found in cell membranes together with proteins and they are also stored as energy sources within certain cells. In plants they are laid down as reserve material for the developing seed, e.g. maize, olives, peanuts, and man obtains his food from these, e.g. maize oil, olive oil, peanut butter. Man and other mammals store fat in adipose tissue under the skin where it forms an insulating layer, and also around the kidneys (as "suet") and other organs where it is a reserve material remaining until the body requires it.

Fats take up the dyes Sudan III and Sudan IV and these dyes are used as positive tests for their presence; fats also give a characteristic translucent effect when rubbed on paper. When fats are digested they are broken down to fatty acids and glycerol.

Proteins

Proteins are used by the body to build new tissues. Protein molecules consist of atoms of carbon, hydrogen and oxygen together with nitrogen atoms; often sulphur and phosphorus atoms are also present. Proteins are essential for body-building and repair because when they are broken down by digestion they release their constituent amino acids which are the "building blocks" for making the body's own proteins. There are over twenty amino acids which are found in the human body, eight of these cannot be synthesised by man and are said to be *essential amino acids*; they must be obtained from food. Plant and animal tissues are the normal sources of proteins. Animal protein is called first class protein since it contains all the essential amino acids and is found in foods such as meat, fish, eggs, milk, cheese. Plant protein is second class protein because it may lack one or two essential amino acids, and is found in peas, beans, lentils, and pulses, etc. Protein gives a pink precipitate when boiled with Millon's reagent.

ENERGY AND THE BASAL METABOLIC RATE

The energy for the body's activities comes from food, but the human body, however, can only convert approximately 15 per cent of the energy of food into mechanical working, the rest being dissipated as heat. The amount of energy required to maintain the functions of the body when you are lying still, warm and without food is called the *basal metabolic rate*.

Metabolism

The sum of all the chemical reactions that take place in the living body is called metabolism, and it is affected by a number of factors.

Exercise

Exercise involves muscular activity which uses energy, and therefore exercise increases the metabolic rate.

Body weight and surface area

The thin person will use less energy when, say, climbing stairs, than a fat person with more body mass to raise against gravity. The surface area of the fat person will be greater than that of the thin person so the fat person will lose more heat energy from the body surface. Therefore, the metabolic rate of the obese person will need to be higher than that of the thin person in order to balance the higher energy expenditure of the heavier body during normal activity.

State of health

Some diseases will slow down the metabolic rate, most commonly those which affect the functioning of the thyroid gland.

Food

The basal metabolic rate is raised after a meal because of energy required by the cells of the gut to churn and move the food in the gut and to produce enzymes etc.

Sexual differences

The basal metabolic rate of women is lower than that of men. For a man it is about 167 420 $J/m^2/h$ and for a woman 154 864 $J/m^2/h$.

Starvation

Starvation reduces the basal metabolic rate considerably due to the lack of energy available for body activity.

Sleep
Sleep reduces the metabolic rate due to the reduction in body activity.

Climate
A cold climate will increase the basal metabolic rate because more energy is required to maintain the body temperature whereas warmer climates decrease the basal metabolic rate because less energy is required to maintain the body temperature.

Age
The basal metabolic rate varies throughout life. At birth it is low; it then rises rapidly during the first year of life and throughout childhood. This is because the child is growing and requires more energy to build up amino acids into the proteins which make up the cells of the body. The basal metabolic rate increases during the teens and then, as the person stops growing, it drops off quite rapidly until it reaches the adult level. It remains at this level throughout adult life until the age of forty when it begins to slow down, until by the age of seventy years the basal metabolic rate is about 70 per cent of what it was at the age of twenty years.

THE FOOD REQUIREMENTS OF THE INDIVIDUAL

The amount and type of food which provides a good diet for a person depends on the age and activity of the person. A person's diet, as mentioned above, must contain carbohydrates, fats and proteins, water, mineral salts, vitamins and roughage, but in the correct proportions. This proportion will vary according to circumstances, for example a child will require a higher proportion of proteins in order to provide the amino acids required to build up the body proteins, so that at least 20 per cent of the child's food intake must be protein. Protein deficiency as is seen in many children of the third world, results in *kwashiorkor* and it is characterised by emaciation, retarded growth and physical weakness. Old people require a good supply of protein in their diet in order to make good the repair of ageing tissues. A healthy man similarly requires 15 per cent protein in his diet in order to replenish worn out and damaged tissues.

The carbohydrate requirements, however, are more related

to the person's occupation. For example, a coalminer who is doing manual work will require more energy producing food than, say, a secretary sitting at a desk all day, because the energy requirements of an individual depend on his basal metabolic rate plus his activity, e.g. sitting requires 63 kJ/h, walking slowly 483 kJ/h, walking upstairs 4200 kJ/h. A man in a sedentary occupation may require 9000 kJ, a man in a moderately active occupation would require about 13 000 kJ and someone in a very active occupation could require 21 000 kJ. This energy requirement is obtained from energy-rich food, i.e. carbohydrates, fats and proteins. (The amount of energy present in food is found experimentally using a bomb calorimeter in which a known weight of food material is burned in oxygen and the amount of heat produced by the food is calculated by measuring the rise in temperature it produces in a given volume of water present in the surrounding water jacket.)

1 g of carbohydrate yields	17.2 kJ
1 g of protein yields	22.2 kJ
1 g of fat yields	38.5 kJ

A manual worker requires a diet containing a higher proportion of carbohydrate than a sedentary worker. An Eskimo requires a diet with a high fat content to provide both the energy required to keep the body warm in spite of the low temperatures, and a layer of fat under the skin to insulate the body.

If a diet contains insufficient food the result is *malnutrition*. If a diet contains an excess of food which is above the requirements of the individual, the excess fat and carbohydrate is laid down as fat around the body and the person becomes obese. This condition can lead to heart trouble and other associated conditions. The simple answer to a weight problem is to cut down on the number of calories (joules) taken in; this does not mean cutting out food completely or cutting out complete meals only to take in a large calorie intake later on in the day, but it does mean eating a sensible calorie-controlled diet which contains minerals, vitamins, roughage, protein, carbohydrates and fats in amounts best suited for the proper functioning of all the parts of the body while enabling it to mobilise its reserve stocks of fuel and use them and thus effect slimming. In other words, if necessary, eat less of every meal.

Vitamins in the diet

Vitamin A

This is fat soluble and is present in the liver oils of fish (e.g. cod) and mammals. Vitamin A is derived from carotene, the yellow pigment present in plants; carotene is absorbed from the intestine and converted into vitamin A in the liver. Vitamin A is also present in certain vegetables, e.g. carrots, lettuce and dark green vegetables, in certain fruits, e.g. apricots, tomatoes, and in dairy produce, e.g. milk, cream, butter and eggs.

Deficiency of this vitamin causes dryness and roughness of the skin, the membranes covering the conjunctiva become dry and disorders of the nervous system can occur. Vitamin A is required for the light-sensitive visual purple or rhodopsin of the retina as visual purple is a derivative of vitamin A. Lack of vitamin A results in night-blindness and degenerative conditions of the nerves and nerve tracts of the central nervous system. If the deficiencies occur early in life there can be retardation of growth together with deformities in bone.

Vitamin B

This is a complex mixture of substances consisting of:

(a) vitamin B_1 (thiamine);
(b) vitamin B_2 (riboflavine);
(c) nicotinic acid;
(d) pyridoxine, B_6;
(e) folic acid;
(f) antipernicious anaemia factor, B_{12};
(g) inositol;
(h) biotin;
(i) para-amino-benzoic acid;
(j) choline.

All these substances are water-soluble.

Vitamin B_1 or thiamine is the anti-neuritic vitamin essential for the normal functioning of the nervous system. It aids the normal metabolism of carbohydrates in the body. Lack of it can lead to paralysis together with oedema and dilation of the heart. The disease is called beriberi and it is common in countries where polished rise is the staple diet, for the husks contain the vitamin and they are cleaned off in the polishing process. Thiamine is found, then, in whole grain cereals, and also in milk, liver and kidneys.

Riboflavin is found in milk, liver, kidney, lean meat and in green vegatables. It is also present in yeast extract and beer. It is slightly soluble in water and can be leached out during the cooking of green vegatables. It is destroyed by sunlight and so milk should not be left in sunlight. Absence of riboflavin from the diet results in a growth of fine blood vessels into the cornea which is normally bloodless and transparent. The growth results in the eyes becoming extra-sensitive to light. The riboflavin deficiency also causes inflamed areas around the angles of the mouth.

Nicotinic acid or niacin is found in yeast, wheat germ, soya beans and peanuts. Deficiency of it causes the disease called pellagra which is characterised by gastric and intestinal disorders, mental symptoms and inflamed patches in the skin. This disease is endemic in countries where the staple diet is maize, which lacks this factor.

Vitamin B_{12} is widely distributed in animal foods, and liver is particularly rich in it. It is called the haematinic principle and it is effective in alleviating the symptoms of pernicious anaemia. Pernicious anaemia is a severe blood disease due to defective functioning of the bone marrow. The average size of the erythrocytes is greater than normal and they have a short life with abnormalities in their metabolism so that the number of erythrocytes in the blood is reduced and the patient suffers from lack of oxygen.

Vitamin C

This is ascorbic acid and is soluble in water. This vitamin is present in green vegetables and turnips, and fruits such as lemons, limes, oranges, apples, tomatoes, blackcurrants and rose-hips. Deficiency of this vitamin causes the disease called scurvy which is characterised by haemorrhages of the gums, mucous membranes, bones and joints and it causes the teeth to fall out. This disease commonly showed itself in sailors in the old days who went for long sea voyages on diets of salt meat and ships' biscuits but no fresh fruit; later, to counteract the effects, limes were issued to all ships of the British Navy.

Vitamin D

This, the anti-rachitic vitamin, is fat soluble. It is the only vitamin than can be produced by the cells in our own bodies; it is produced in the adipose tissue under the skin by the action of the ultra-violet light from the sun. Other sources are fatty fish, milk, cheese made in summer, egg yolk and margarine which has vitamin D added to it during manufacture. Deficiency of this vitamin causes rickets in children, a disease in which the bones do not possess normal rigidity and strength since they contain less than the required amount of calcium and phosphorus because vitamin D affects the uptake of calcium from the intestine (*see* Chapter 7). If there is a deficiency of vitamin D in adults, calcium is taken out of the bones by the blood to offset the lack of it resulting from deficient uptake. This disease is called osteomalacia; the calcium is not replenished and the strength of the bones is diminished.

Vitamin E

This is the name given to a group of fat-soluble substances occurring naturally in foodstuffs such as seed embryos, milk and eggs. Opinion varies as to its significance as a vitamin in man's diet. It has been shown to prevent abortion in rat embryos and is often referred to as the antisterility vitamin, as lack of it causes sterility in the male.

Vitamin K

This is a name given to a number of compounds which all seem to have the same vitamin activity; some are fat soluble and others show a tendency to water solubility. Vitamin K is concerned with the formation of prothrombin which is required for the clotting of blood. Uptake of vitamin K from the gut is dependent on the presence of bile and if, for any reason, the bile duct is obstructed, vitamin K deficiency symptoms occur.

Minerals in the diet

The body needs certain minerals to provide raw materials for the proper growth and functioning of the body. Plants get their minerals from the soil water and so water is the indirect source of minerals in our diet. The term "mineral" refers to all chemical elements other than carbon, hydrogen, oxygen and nitrogen. There are about twenty elements which are required in appreciable amounts by the body; these include (in decreasing quantities) calcium, phosphorus potassium, sulphur, sodium, chlorine, magnesium, copper and iodine. Some other elements are required in trace amounts only; these are known as the *micronutrients* and include cobalt, selenium, molybdenum, zinc, silicon, aluminium, chromium, arsenic, boron, fluorine and nickel. Micronutrients are essential for normal growth and development of the body.

Calcium

Calcium, together with phosphorus and magnesium, is required for the formation of bones and teeth, and calcium contributes about 1 kg to the weight of a 70 kg adult male. Besides being a constituent of bones, calcium is essential for the proper functioning of the nerves and for the clotting of the blood. The ratio of calcium to phosphorus in the diet is critical. There must be 1.5 parts of phosphorus in the diet of an adult to 1 part calcium in order that the calcium is absorbed correctly. If there is too much phosphorus or too little in proportion to the amount of calcium in the diet, supplies of either calcium or phosphorus can be withdrawn from the bones. If the ratio of phosphorus to calcium becomes 6:1, rickets may develop. A good supply of vitamin D in the diet can, to some extent, compensate for the wrong ratio. Children require a good supply of calcium in their diet to enable their bodies to build up their skeleton, as do pregnant women to meet the requirements of the foetus and also, at the end of pregnancy, for lactation.

Calcium is present in milk and milk products, and is added to flour so that the population will not suffer from a deficiency of it.

Iron

Iron is needed for the manufacture of haemoglobin which carries oxygen around the body in the erythrocytes or red corpuscles. Myoglobin is another iron-containing red pigment which is present in muscles and responsible for the transport of oxygen. Oxygen is used within the cells for energy release and some of the enzymes used in this process, including cytochrome, contain iron. The rest of the iron in the body is stored as ferritin in the liver, spleen and red bone marrow. If there is insufficient iron in the diet the condition known as *anaemia* is produced. Anaemic persons have blood which is pale in colour and they feel tired even though they have had sufficient sleep. The reason for this is that the red cells are not able to supply enough oxygen to the muscle cells to oxidise food completely and produce the correct amount of energy. Liver, meat, black sausage and eggs are all good sources of iron as are dark green leaves of vegetables, black treacle, dried fruit and curry powder.

Phosphorus

Phosphorus, together with calcium, is a constituent of bones and teeth. It is also present in tissue fluids and the body maintains a balance between that phosphorus in the bone and that in the tissue fluid. Phosphorus plays an essential part in energy release in the adenosine diphosphate/adenosine triphosphate cycle.

Magnesium

Magnesium is also an important component of bone and teeth and is important for normal metabolism. It is present in all types of food and is widely found in green vegetables since it is a component of chlorophyll.

Sodium

Sodium is present in all body fluids and shortage of sodium in the body results in cramps. Sodium is lost in sweat and urine and has to be replaced from the food in the diet. In conditions of extreme heat and exercise, symptoms of sodium deficiency or heat fatigue occur; these are muscular weakness, drowsiness and mental confusion. These symptoms can be alleviated by drinking salt water. This condition often occurs in miners and athletes but can occur in others in very hot, dry conditions. Sodium is present in bacon, cornflakes, cheese, bread, butter, eggs and fish.

Chlorine

Chlorine is present in chloride ions throughout all the body tissues. It is found combined with hydrogen as hydrochloric acid in the gastric juice.

Iodine

Iodine is contained in thyroxine, a hormone produced by the thyroid gland, and deficiency of iodine in the diet produces enlargement of the thyroid gland which is known as *simple goitre*. Thyroxine controls the metabolic rate. Iodine is found in vegetables grown in soil containing iodine, sea fish, shellfish and iodised table salt.

SELF-ASSESSMENT QUESTIONS

1. Proteins are built up of smaller units called:
   (a) monosaccharides;
   (b) fatty acids;
   (c) glycerol;
   (d) amino acids;
   (e) disaccharides.

2. Which of the following is a polysaccharide?
   (a) cellulose;
   (b) sucrose;
   (c) glucose;
   (d) maltose;
   (e) fructose.

3. What does protein deficiency in the diet cause?
   (a) scurvy;
   (b) kwashiorkor;
   (c) rickets;
   (d) osteomalacia;
   (e) night-blindness.

4. One of the following foods supplies the mineral element needed for the healthy functioning of the thyroid gland. Which is it?
   (a) oranges;
   (b) lemons;
   (c) unpolished rice;
   (d) milk;
   (e) sea fish.

5. Select a food that can provide roughage for the diet:
   (a) cream;
   (b) bananas;
   (c) roast beef;
   (d) cabbage;
   (e) mutton chops.

6. Which person requires the smallest number of kilojoules in their diet?
   (a) a teenage typist;
   (b) a housewife;
   (c) a newborn baby;
   (d) a mineworker;
   (e) an old age pensioner.

7. Which of the following vitamins is fat soluble?
   (a) vitamin A;
   (b) vitamin B_1;
   (c) vitamin B_2;
   (d) vitamin C;
   (e) nicotinic acid.

ASSIGNMENTS

1. List all the food you take in for a period of a week. Include the amounts of each food constituent together with its calorific value. Put your data in a chart and examine your diet from the point of
   (a) calorific content;
   (b) vitamin content;
   (c) nutrient content.

2. Is your diet adequate for your needs? Write a short account of your diet and its suitability for your way of life.

CUMULATIVE QUESTIONS, CHAPTERS 1–5

1. The smallest blood vessel in the body is:
   (a) a capillary;
   (b) a vein;
   (c) an artery;
   (d) a red corpuscle;
   (e) a coronary vessel.

2. The largest artery in the body is the:
   (a) inferior vena cava;
   (b) capillary;
   (c) ventricle;
   (d) aorta;
   (e) hepatic portal.

3. The instrument used to measure the blood-pressure is:
   (a) a watch;
   (b) an arm band;
   (c) a sphygmomanometer;
   (d) a spirometer;
   (e) an electrocardiograph.

4. In tissue respiration oxygen reacts with glucose to produce:

(a) energy and water;
(b) energy, lactic acid and carbon dioxide;
(c) carbon dioxide and water;
(d) carbon dioxide, energy and water;
(e) heat and water.

5. When you run to catch a bus you continue to breathe hard for some minutes afterwards because:

(a) your pulse is higher and takes time to slow down;
(b) it takes time for you to slow down your breathing rate;
(c) you are hot and tired;
(d) you need to get rid of the carbon dioxide produced when you run;
(e) you are paying off the oxygen debt acquired while you were running.

6. The breathing rate is regulated by the brain:

(a) principally according to the level of oxygen in the blood;
(b) according to the rate of the pulse;
(c) according to the blood pressure in the arteries;
(d) principally according to the carbon dioxide content of the blood;
(e) under voluntary control.

7. The approximate percentage of carbon dioxide in the atmosphere is:

(a) 21;   (d) 0.04;
(b) 0.4;   (e) 24.
(c) 4;

8. The function of respiration is to:

(a) produce heat;
(b) produce carbon dioxide;
(c) produce energy;
(d) take in oxygen from the air;
(e) breathe so that you might live.

9. Lymph flows in thin-walled vessels called:

(a) lymphocytes;   (e) lymph glands.
(b) capillaries;
(c) lymphatics;
(d) arterioles;

Digestion

After studying this chapter you should be able to:
* identify the various tissues found in the gut;
* understand the functions of the various parts of the gut;
* relate the structure of the tissues of the gut with their functions.

The alimentary canal is a tube commencing at the mouth and terminating at the anus. It is differentiated into the mouth, pharynx, oesophagus, stomach, small intestine and large intestine (*see* Figs. 41, 42 and 50).

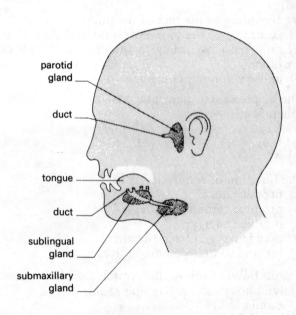

Fig. 41. *The salivary glands.*

THE MOUTH

The histology of the mouth

The mouth is lined with mucous membrane which is composed of stratified epithelium penetrated by vascular papillae and the buccal glands (small glands which secrete mucus to lubricate the mouth).

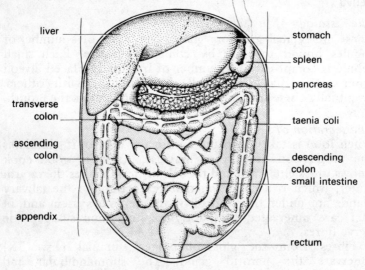

Fig. 42. *Alimentary canal in the abdomen of man.*

The functions of the mouth

Food enters the gut through the mouth, where it is moistened by saliva, chewed by the teeth and churned by the tongue.

Digestive enzyme action results in breaking large food molecules into smaller units which can be absorbed into the bloodstream. The enzymes must be in direct contact with the food molecules, so the greater the surface area the food has the more food molecules are in contact with the enzymes, making faster digestion possible. Chewing is the mechanical breakdown of large chunks of food into smaller food particles producing a larger surface area. It prepares food for digestion by enzymes. Thorough chewing mixes the food with saliva which is produced by three pairs of salivary glands (*see* Fig. 41). In man these are the parotids, submaxillary and sublingual glands. Saliva contains salivary amylase, an enzyme which works best in approximately neutral conditions (pH 6.7). The saliva contains buffers which help maintain this pH. Salivary amylase converts starch into maltose. Saliva moistens and lubricates food and helps to make it into a ball or bolus to be swallowed. *Saliva is a thick colourless liquid that contains water, mucin, salts*

Saliva *& the dig enzyme salivary amylase.*

The histology of the salivary glands

These are typical secretory glands composed of a number of lobules bound together by connective tissue. Each small lobule is composed of a number of irregularly placed alveoli from which passes a small duct, uniting with others. Eventually these link into one duct which enters the mouth.

The secretion of saliva

When food is taken into the mouth the taste of it stimulates the production of salivary amylase. As every good cook knows, the sight, smell and thought of food makes the mouth water, which is a conditioned reflex action, so the salivary glands are under the control of the nervous system and, in fact, are innervated by parasympathetic and sympathetic nerve fibres.

Fibres from the glossopharyngeal (cranial nerve IX) innervate the parotid gland. The submandibular and sublingual glands receive their innervation from the facial nerve (cranial nerve VII).

The functions of saliva

The salivary glands contain two types of glandular cells:

(a) mucous, which produce a viscous secretion containing mucin;

(b) serous, producing a watery secretion containing salivary amylase or ptyalin.

Saliva passes into the mouth where it:

(a) moistens dry food;
(b) dissolves the food and aids taste;
(c) helps the breakdown of starch to maltose at an optimum pH of 6.7 because it contains the enzyme salivary amylase;
(d) lubricates the food particles because it contains mucin;
(e) helps speech.

The teeth

Amphibians and reptiles have teeth which are simple cone shaped structures, i.e. they are *homodonts*. Mammals have a variety of teeth which are modified for various purposes, i.e. they are *heterodonts*.

The teeth are specialised as follows:

(a) incisors, found at the front of the jaw and used for nipping off small portions of food;

(b) canines, pointed teeth used for piercing, tearing and biting food;

(c) premolars and molars, found at the back of the mouth where they grind food reducing it to small particles that can be swallowed.

There are two sets of teeth in man: the milk, or temporary, set of teeth which do not include molars and are shed early in life; and the permanent dentition which follows the milk teeth. Since there are these two sets of teeth the term *diphyodont* is used to describe this condition.

Histology of teeth

Teeth are composed of three calcified tissues: the outermost layer of hard enamel, covering the crown; the main substance of the tooth, dentine, which is a hard dense bone-like substance which does not contain Haversian canals or cells but is pierced by fine tubules or canaliculi which radiate out from the central pulp cavity; and a soft jelly-like connective tissue containing blood vessels, lymph vessels and many nerve fibres.

The root of the tooth is covered with cement which is a thin layer of lamellated bone containing lacunae and canaliculi but no Haversian canals. The cement is covered with periosteum which also lines the socket (*see* Fig. 43).

Types of teeth

Mammals, unlike other vertebrates, have two sets of teeth (diphyodont), with a notable exception of the anteaters which do not possess teeth.

From the dietary point of view there are three groups of animals:

(a) herbivores which feed on plants;

(b) carnivores which feed on the flesh of animals;

(c) omnivores which feed on a mixed diet of plant and animal material.

These divisions are reflected in the type of teeth present (*see* Fig. 44).

The number of types of teeth in half of the upper jaw and in half of the lower jaw are expressed in the dental formula.

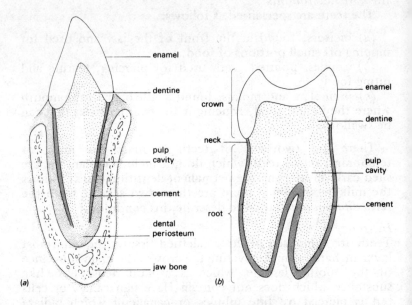

Fig. 43. *(a) V.S. incisor. (b) V.S. molar.*

Man, who is an omnivore because he feeds on both animal and plant material, has two incisors on top and two below, one canine above and one below, two premolars above and two below and three molars above and three below. Man's dental formula is expressed as:

$$i\ \frac{2}{2},\ c\ \frac{1}{1},\ pm\ \frac{2}{2},\ m\ \frac{3}{3}.$$

Gnawing animals have bladelike incisors for grazing. Rodents which are herbivores have enamel only on the anterior surface of the incisors. They grow continuously and this growth keeps pace with wear. Rabbits have an additional small pair of incisors behind their first pair. The crowns of their incisors are covered with enamel. Vampire bats have small bladelike incisors which cut skin like lancets allowing the animal to suck blood. The incisors of elephants are modified into tusks which can be used to move trees.

Canines are well developed in carnivores and the canines are used for killing and tearing the flesh of prey. Canine teeth also are often specialised for purposes other than that directly concerned with feeding, for example, wild boars have canines which are tusks, which they use for defensive and aggressive behaviour. Walruses also have large canines developed as tusks, which they use for hauling their huge bodies out of the water and also for levering out shellfish. Rodents have no canine teeth. The gap which is left is called the *diastema*. In the Cetacea, the Narwhal male has only one canine on the left side of the upper jaw. It is very long, sometimes several metres, and points straight out in front of the animal.

Herbivores have premolars and molars with broad crowns. The enamel is folded into the dentine and chewing wears the surface producing a series of ridges as the soft dentine is worn away faster than the hard enamel. These ridges help animals like cows and horses to chew the plant material which forms their food.

Carnivores have blade-like cutting edges to their molars and premolars. The last premolar of the upper jaw and the

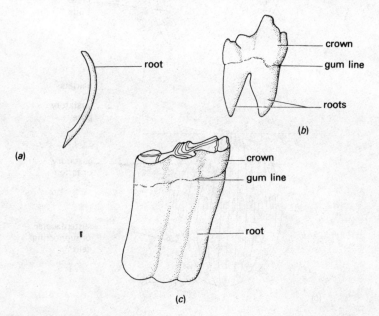

Fig. 44. *Various types of teeth: (a) rodent incisor; (b) carnassial; (c) horse molar.*

first molar of the lower jaw form sharp cutting edges with each other. These are the *carnassial teeth*.

Porpoises and dolphins are unusual in that their teeth are peglike and more or less identical, i.e. they are homodont, since they are mostly only used to catch fish which are then swallowed whole.

The tongue

This is attached to the posterior ventral wall of the mouth or buccal cavity. It is composed of striated muscle whose muscle fibres run in three different planes, i.e. longitudinal, transverse and vertical. It is covered by mucous membrane with stratified epithelium. The upper surface is covered with lingual papillae which give it a rough appearance. There are three kinds of lingual papillae.

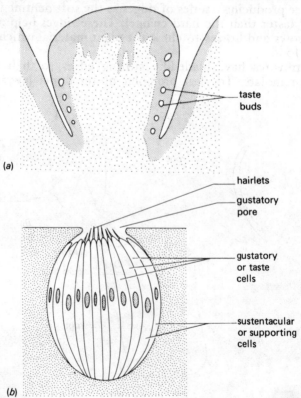

Fig. 45. *(a) Section of circumvallate papilla showing taste buds. (b) A taste bud.*

(a) Circumvallate, *see* Fig. 45. There are about twelve or thirteen of these comparatively large circular projections situated in a V-shaped line, the apex of the V points towards the back of the tongue. In man, taste buds cover the sides of each papilla and the vallum surrounding it. Innervation is by the glossopharyngeal nerve.

(b) Filiform or conical papillae. These are found covering the rest of the tongue. In cats they are curved, claw-shaped, hard and horny and help scrape meat from bones.

(c) Fungiform. These larger papillae are scattered amongst the filiform papillae. They are highly vascular projections, redder than their fellows and lying embedded in depressions in the membrane. Taste buds and mucus-producing lingual glands are present in their epithelium.

A healthy tongue is pink but in ill health the lingual mucosa is not shed and remains together with bacteria producing a furred appearance. The tongue aids speech and helps the teeth in chewing by churning the food around the mouth.

THE PHARYNX

Structure and functions

The movements of the tongue push the food into the pharynx which is the cavity situated behind the soft palate. The pharynx is lined by stratified epithelium and kept moist by many mucous glands; this layer is attached by areolar connective tissue to an underlying fibrous membrane encircled by striated constrictor muscles whose action contributes to swallowing. Two internal nares enter the pharynx dorsally, laterally there are two Eustachian tubes linking the pharynx to the middle ear, used for equalising the air pressure on the ear drums (or tympanic membranes).

The opening from the pharynx into the larynx (or voice-box) is called the glottis: it is closed by the valve-like epiglottis which prevents food entering the trachea (or windpipe).

The oesophagus conveys the food from the pharynx to the stomach and when swallowing occurs reflex action closes all openings to the pharynx except the entrance to the oesophagus; the pharynx muscles contract and the round mass of moist food formed in the mouth, i.e. the bolus, drops into the oesophagus.

THE OESOPHAGUS

Histology of the oesophagus

Mucous membrane forms its inner surface and the lining, stratified epithelium, is perforated by many mucus-secreting glands. The mucous membrane is attached by areolar connective tissue to the underlying unstriated muscle fibre coat called the muscularis mucosae, this layer being attached by areolar connective tissue, containing blood vessels and lymphatics, directly to the main muscle layers; the inner circular layer and the outer longitudinal layer. In the upper third of the oesophagus, these muscles are voluntary but posteriorly they are involuntary (*see* Fig. 46).

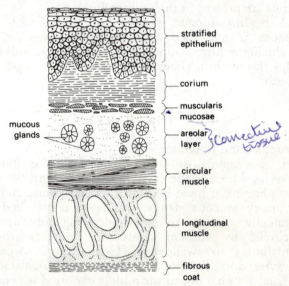

stratified epithelium

corium

muscularis mucosae

mucous glands

areolar layer — connective tissue

circular muscle

longitudinal muscle

fibrous coat

Fig. 46. *Section through the wall of the human oesophagus.*

Functions of the oesophagus

Swallowing is a voluntary process but once the bolus passes into the oesophagus, the involuntary muscles propel it towards the stomach by means of alternating contractions of the circular and longitudinal muscles. Their action produces a series of waves of contraction and relaxation in the walls of the canal; this action is known as *peristalsis*. The oesophagus does not produce enzymes, it is simply a canal carrying the food to the stomach.

THE STOMACH

Histology of the stomach

The stomach is lined by the *mucous coat* of columnar epithelium whose goblet cells produce mucus continuously to lubricate the contents; the lining is perforated by the tubular gastric glands which extend to the *muscularis mucosae* (an inner circular and an outer longitudinal layer of non-striated muscle fibres — *see* Fig. 47).

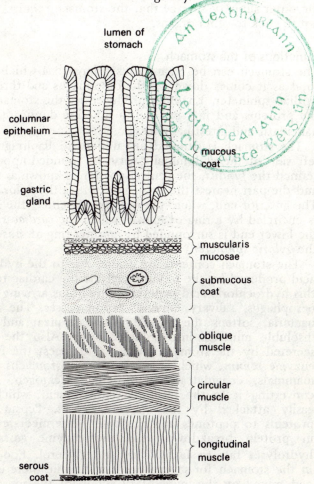

Fig. 47. *Section through the wall of the stomach.*

A *submucous coat* of areolar connective tissue, containing blood vessels and lymphatics and a gangliated nerve plexus, binds the mucous coat to the main *muscle coat*. Three layers of involuntary muscle make up the muscle coat and work together to churn food and gastric juice. The inner layer has oblique fibres, the middle layer circular fibres and the outer layer longitudinal muscle fibres. The last two layers have a nerve plexus running between them.

The *serous coat* derived from the peritoneum provides the outer slippery surface that the stomach requires in order not to stick to other organs.

Functions of the stomach

The stomach can be regarded as a large bag which receives food as it comes down via the oesophagus and through the cardiac sphincter. Enzymes produced by the stomach digest the proteins and fats in food; a function duplicated by the pancreatic and intestinal juices.

The stomach is situated just under the diaphragm on the left side of the abdominal cavity. Its rounded upper part is termed the *fundus*, the central portion is known as the *body* and the part nearest the small intestine is the *pyloric region*. The anterior end, where the oesophagus enters the stomach, is encircled by a ring of muscle called the *cardiac sphincter*, the lower end is surrounded by another ring of muscle called the *pyloric sphincter*.

The stomach contents are acidic, due to the hydrochloric acid produced by the oxyntic cells of the cardiac region. As this hydrochloric acid penetrates the bolus arriving from the oesophagus, salivary amylase action ceases. The acid kills bacteria, softens the muscle fibres of meat and converts insoluble minerals into soluble forms. Also the enzymes secreted by the stomach walls act to digest the food. The enzyme *rennin*, which is present in the stomachs of young mammals, coagulates the protein, caseinogen, of milk, converting it to the insoluble curd, casein, which can be easily attacked by the enzyme *pepsin*. Pepsin converts proteins to peptones and proteoses (intermediate products in protein breakdown) and the enzyme *gastric lipase* hydrolyses fats to fatty acids and glycerol. Food remains in the stomach for several hours and while there is churned and mixed by the involuntary action of the muscle in the wall of the stomach.

The pyloric sphincter controls the flow of partially-digested food out of the stomach: it relaxes when food in the stomach is sufficiently fluid, allowing it to enter the first part of the small intestine, the duodenum.

Secretion of gastric juices

Flow of gastric juice is controlled by parasympathetic fibres from the vagus nerves and sympathetic fibres from the coeliac plexus. The presence of food in the stomach is the main stimulus for the secretion of the gastric juice. However, sight, smell and taste of food play their part in stimulating

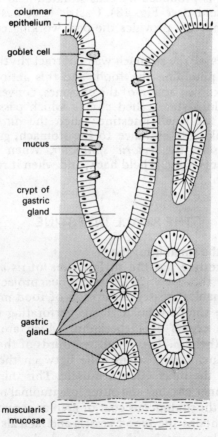

columnar epithelium

goblet cell

mucus

crypt of gastric gland

gastric gland

muscularis mucosae

Fig. 48. *Section through the fundus region of the stomach showing detail of gastric glands.*

the vagus nerve (cranial nerve X) whose nerve endings release
acetyl choline which stimulates the secretory cells to produce
a flow of gastric juice. The stimulus of the presence of
proteins in the stomach ensures that the juice is rich in
pepsin and that the walls of the gastric mucosa in the pyloric
region secrete the hormone *gastrin*. Gastrin circulates in the
blood and stimulates the gastric glands to secrete more gastric
juice. This juice contains the enzymes rennin, pepsin and
gastric lipase. These enzymes are secreted by the peptic cells
in the wall of the gastric glands. Deeper regions of the gastric
glands secrete hydrochloric acid. Gastric juice also contains
mucus, which is produced by cells situated near the neck of
the gastric glands (*see* Fig. 48). Gastric juice has a pH value
less than 2 and this provides the best working environment
for pepsin.

The muscles of the stomach wall contract rhythmically, so
mechanically pounding the food, and this action, together
with the breakdown action of the enzymes, converts the food
into a semi-fluid state called *chyme* which passes into the
first part of the small intestine called the *duodenum*. Its
entrance is, like the entrance to the stomach, guarded by a
sphincter muscle, the *pyloric sphincter*. When this ring of
muscle contracts, food is held back, and when it relaxes, food
passes through.

THE SMALL INTESTINE

Histology of the small intestine

The small intestine has the surface area of its *mucous coat*
increased by the presence of numerous projections called
villi, which enables greater absorption of food materials (*see*
Fig. 49). The simple tubular glands perforating this mucous
coat are the crypts of Lieberkühn, which are lined with
columnar epithelium. The distal two-thirds of the duodenum
contain Brünner's glands which lie between the muscularis
mucosae and the inner circular muscle. This thin muscularis
mucosae of inner circular and outer longitudinal muscle fibres
is adjacent to the submucous coat of areolar connective tissue.
This layer contains the blood vessels and lacteals which serve
the villi. It also contains the gangliated plexus of *Meissner*,
whose fibres innervate the muscle fibres of the muscularis

mucosae and the villi and glands. The involuntary muscular coat has two layers, a thick inner band of circular fibres and an outer band of longitudinal fibres. Between the two there are lymphatics and a gangliated plexus of myelinate nerve fibres known as *Auerbach's plexus* controlling the peristaltic action. The serous layer provides the small intestine with a smooth outer coat.

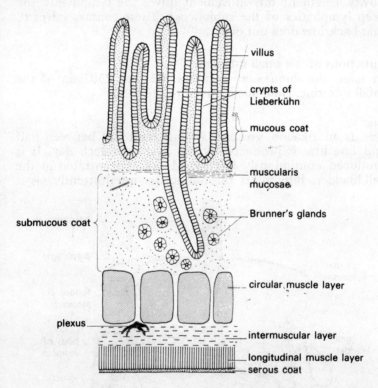

villus

crypts of Lieberkühn

mucous coat

muscularis mucosae

Brunner's glands

submucous coat

circular muscle layer

plexus

intermuscular layer

longitudinal muscle layer
serous coat

Fig. 49. *Section through the wall of the small intestine.*

Villi

As previously mentioned, the inner surface of the small intestine is covered by an enormous number of projections called villi, which increase the surface area of this part of the gut where absorption of digested food material takes place. Each villus contains a venule, an arteriole, and a blind-ended lacteal (lymph vessel).

Monosaccharides and amino acids pass through the surface of the villus and into the venules which connect with the hepatic portal vein which, in turn, takes them to the liver. Fatty acids and glycerol are absorbed by the lacteals which pass them into the lymphatic system where they eventually enter the general circulation through the thoracic duct. The villi are constantly moving, they shorten quickly and then slowly lengthen; this movement drives the lymph into the deep lymphatics of the submucosa. These contain valves so that backflow does not occur.

Functions of the small intestine
In man, the *duodenum* occupies the first 250 mm of the small intestine. Here bile is added to the food.

Bile
Bile is an alkaline, greenish viscous fluid and between half and one litre is produced by the liver cells each day. It is produced continuously but stored and concentrated in the gall-bladder. Bile enters the duodenum intermittently, as it

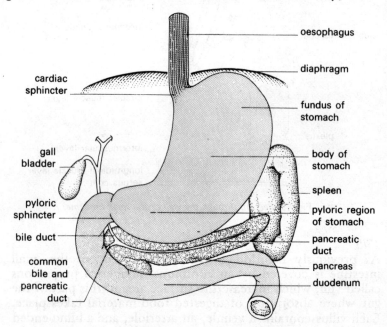

Fig. 50. *The pancreas and the bile and pancreatic ducts.*

is required. It is an aqueous solution of the sodium salts of glycocholic and taurocholic acids, mucin, cholesterol, fats and inorganic salts. In addition there are two bile pigments, the reddish bilirubin and the greenish biliverdin, which are formed from the hæmatin of the hæmoglobin released during the breakdown of the erythrocytes of the blood. Bile is a true excretion which is used by the body in the digestion of fats. The bile salts activate lipase and emulsify fats thus aiding digestion, because of the enormous total surface area of the fat droplets in an emulsion.

Histology of the gall-bladder
This bag-like structure which stores bile is lined by columnar epithelium. Its coat is fibrous, muscular and elastic. The epithelial cells are capable of absorbing water and therefore concentrate bile while it is in the gall-bladder.

In man the bile duct leading from the gall-bladder unites with one leading from the pancreas forming the *common bile and pancreatic duct* which enters the duodenum (*see* Fig. 50).

Hormonal control of bile secretion
Bile is secreted continuously and this is independent of nervous stimuli. The presence of food in the duodenum causes the walls of the duodenum to produce the hormone *cholecystokinin,* which stimulates intermittent emptying of the gall-bladder contents on to the food.

The pancreas
Histology of the pancreas
This racemose gland has long tubular *alveoli* and loose areolar connective tissue. It contains special cell masses called the *islets of Langerhans* (endocrine organs which produce the hormone insulin).

This is the internal secretion of the pancreas. The external secretion is the pancreatic juice, containing enzymes, which passes along the ducts leading from the alveoli and enter the duodenum through the common bile and pancreatic duct.

Pancreatic secretions
The pancreas produces two secretions:

(*a*) an internal secretion, the hormone insulin, which is sent direct into the blood stream;

(b) an external secretion which is sent into the gut through the pancreatic duct, produced as a result of stimulation by the vagus nerve. This contains the digestive enzymes lipase, amylase and trypsin. Lipase converts fats to fatty acids and glycerol. Amylase converts starch into maltose. Trypsin is secreted as trypinogen and converted into trypsin farther down the intestine by the action of the activator enterokinase. Trypsin converts proteins into polypeptides.

Most of the digestion of food takes place in the small intestine and it remains here longer than in any other part of the alimentary canal.

Hormonal control of the secretion of pancreatic juice
Hydrochloric acid (from the stomach), water, meat juices, bread, alcohol and fats all stimulate the walls of the small intestine to produce the hormone *secretin* which causes the pancreas to secrete a weak solution deficient in enzymes. The presence of hydrochloric acid in the small intestine also stimulates the intestinal walls to produce another hormone *pancreozymin* which causes the production of a very rich secretion of the enzymes lipase, amylase and the precursor trypsinogen from the pancreas.

The succus entericus
The total secretion from the intestine is known as the intestinal juice or succus entericus. Chyme in the intestine causes the production of the juice and it is maintained by the action of the hormone enterocrinin which is produced by the walls of the small intestine. The enzymes of the succus entericus complete digestion. Maltase converts maltose into glucose while lactase acts on lactose producing glucose and fructose; these monosaccharides are then ready to be absorbed. Erepsin completes the breakdown of proteins to form amino acids and the amino acids and monosaccharides are taken to the liver by the hepatic portal vein. Note that the enzymes are produced by cells lining the crypts of Lieberkühn whilst Brünner's glands secrete mucus and alkaline fluid, which neutralises the hydrochloric acid.

Segmentation and peristalsis
The chyme is mixed and moved onwards by the action of the gut. Firstly, it is brought about by segmentation: the circular muscle divides the length of the small intestine into a number of equal-sized segments. A few seconds after this a new

contraction of circular muscle takes place in the middle of each segment and the previously contracted muscle relaxes. Initiation of this action takes place independently of the nervous system, it is *myogenic*, i.e. it takes place in the muscle itself. The *peristalsis* brings about onwards movement of the chyme, the circular muscle contractions behind the mass of chyme pushing it forwards. The waves of contraction travel at a speed of between 20 and 250 mm/s. They pass along the whole length of the small intestine. This action is under direct nervous control, receptors within the mucous membrane responding to the stimulus of distention.

The result of all the activities in the small intestine is to convert the chyme into an emulsion called *chyle* from which the products of digestion are absorbed, i.e. monosaccharides, fatty acids, glycerol and amino acids together with vitamins, inorganic salts and water.

THE LARGE INTESTINE

Histology of the large intestine or colon

The mucous membrane has no villi and consists of many densely packed simple tubular glands which contain lots of goblet cells, producing mucus. At this point in the gut the undigested material is passing towards the rectum and water is being absorbed from it. Therefore, lubrication is of prime importance. The submucosa is composed of areolar connective tissue and the muscle coat consists of a well-developed circular layer of muscle fibres and an outer layer of longitudinal involuntary muscles fibres. The latter are gathered into three thick longitudinal bands called the taenia coli. These produce the puckering in the gut wall. The serous coat is absent near the posterior end of the large intestine.

Function of the large intestine

Here, absorption of water and sodium takes place, undigested food in fluid state has its water removed and faeces are formed. Bacteria and yeast produce vitamin K, vitamin B and amino acids for the body.

THE RECTUM

Faeces are the brownish semi-solid masses passed through the anus. They contain undigested material such as cellulose, fatty acids, sterols and phospholipids. There is also nitrogenous material in the form of some undigested protein, cell debris, bacteria and inorganic material consisting chiefly of calcium and phosphate.

Stercobilin formed from bile pigment gives faeces their colour, the odour is due to skatole and indole formed by bacterial action on the amino acid *tryptophan*. Typical analysis of faeces gives 70 per cent water, 10 per cent dead bacteria and 20 per cent unabsorbed, undigested food residues. Faeces are emptied from the rectum by the reflex-controlled action called *defaecation*.

SELF-ASSESSMENT QUESTIONS

1. The surface area of the small intestine is greatly increased by the presence of:
  (a) the duodenum;
  (b) villi;
  (c) glomeruli;
  (d) alveolar tissue;
  (e) sphincter muscles.

2. A villus in the small intestine has:
  (a) blood capillaries and lacteals;
  (b) ducts from the pancreas;
  (c) enzyme-producing cells;
  (d) enzyme-producing cells and capillaries;
  (e) ducts from the gall-bladder.

3. The oesophagus connects:
  (a) the pharynx and lungs;
  (b) the mouth and lungs;
  (c) the duodenum and the pancreas;
  (d) the gall-bladder and the mouth;
  (e) the mouth and stomach.

4. The digestion of starch begins in the:
  (a) the oesophagus;
  (b) the mouth;
  (c) the large intestine;
  (d) the small intestine;
  (e) the stomach.

5. Bile is secreted by the:
   (a) the gall-bladder;
   (b) the liver;
   (c) the pancreas;
   (d) the duodenum;
   (e) the kidney.

6. When fats are digested in the gut the substances formed are:
   (a) amino acids;
   (b) alkalis;
   (c) reducing sugars;
   (d) fatty acids and glycerol;
   (e) bilirubin and biliverdin.

7. Protein is digested by:
   (a) bile;
   (b) lipase;
   (c) pepsin;
   (d) enterokinase;
   (e) sucrase.

8. Saliva contains the enzyme:
   (a) lipase;
   (b) rennin;
   (c) pepsin;
   (d) salivary amylase;
   (e) erepsin.

ASSIGNMENTS

You are given a brown bread ham sandwich and a glass of milk.

1. Make a large diagram of the alimentary canal of man and show where that sandwich is digested.

2. Describe how the digestion of this meal occurs.

CHAPTER SEVEN

The Liver

CHAPTER OBJECTIVES

After studying this chapter you should be able to:
* describe the structure of the liver and show the relationship that exists between the liver cells, bile ducts and the blood vessels;
* show the importance of the liver in the metabolism of the body;
* show how the liver cells maintain the composition of the blood;
* explain how it acts in a secretory capacity producing bile;
* understand the role the liver plays in the excretory process, deamination of amino acids and production of urea by means of the ornithine cycle.

STRUCTURE OF THE HUMAN LIVER

Gross anatomy

The liver is a large organ situated beneath the diaphragm at the anterior end of the abdominal cavity. It develops as a tubular structure, with its epithelium continuous with the epithelium lining the alimentary canal via the bile duct. The anterior surface of the liver is triangular with the apex pointing towards the left. The hepatic portal vein, hepatic arteries, lymphatics, bile ducts and nerves enter and leave the organ through the portal of the liver which is a deep fissure. A fold of peritoneum, called the falciform ligament, divides the liver into a large right and a smaller left lobe. The right lobe is separated off from two smaller lobes:

(a) the quadrate, situated on the under side of the right lobe and separated from it by the impression of the gall-bladder; and

(b) the caudate or Spigelian lobe which lies on the posterior side (see Fig. 51).

Histological structure

In section it is possible to see that the liver is made up of hexagonal lobules bound together by connective tissue, containing also bile ducts, nerves and interlobular vessels (*see* Fig. 52). The veins are branches of the portal vein and they bring the products of carbohydrate digestion, glucose, and the products of protein digestion, the amino acids, from the gut to the liver.

The arteries are branches of the hepatic artery bringing oxygen to the liver cells. Interlobular bile ducts collect up bile produced by the liver cells and lymphatics carrying lymph are also present between the lobules. Blood passes from the periphery of the liver lobules to the centre where

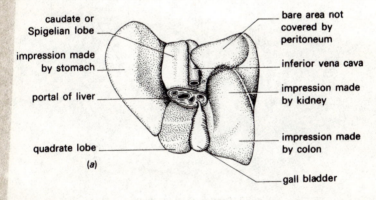

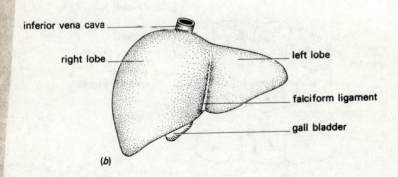

Fig. 51. *(a) Posterior and dorsal surface of liver. (b) Ventral view of liver.*

the intralobular vein takes deoxygenated blood away from the cells; it is a tributary of the hepatic vein.

Liver cells are polyhedral; their coarsely-granular cytoplasm contains glycogen, iron and fatty material depending on the state of activity. The cells possess large, round nuclei and some are binucleate. They radiate in lines, or cords, from the intralobular vessels to the edges of the lobules (*see* Figs. 52 and 53.) Each liver cord consists of one or two rows of liver cells. Between the rows there are tiny channels (called bile

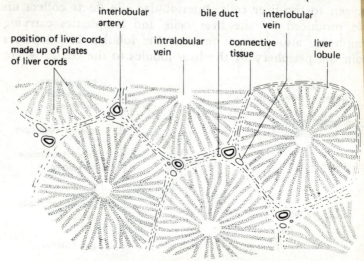

interlobular artery   bile duct   interlobular vein

position of liver cords made up of plates of liver cords   intralobular vein   connective tissue   liver lobule

Fig. 52. *Section through the liver.*

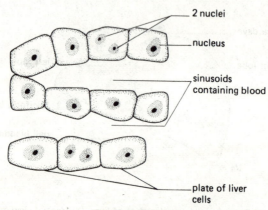

2 nuclei

nucleus

sinusoids containing blood

plate of liver cells

Fig. 53. *Liver cells.*

canaliculi) which drain bile peripherally in a lobule to the bile ducts.

FUNCTIONS OF THE LIVER

The liver is of prime importance in the metabolism of the body and carries out many functions. The most significant ones are the regulation of blood sugar and amino acid levels, described below.

Regulation of the blood sugar level

Shortly after a meal, and while it is being absorbed, blood sugar from the digestion of carbohydrates is carried from the gut to the liver in the hepatic portal vein. The soluble glucose must be converted into insoluble glycogen (animal starch), otherwise it would circulate around the body and upset the osmotic balance between the interstitial fluids and the cells within the body, so the liver helps to maintain a constant level of glucose within the blood — a homeostatic function. Conversion of glucose into glycogen takes place in the liver under the influence of the hormone insulin, which is secreted into the blood by the islets of Langerhans in the pancreas.

Glucose in the liver is also broken down in the process of cell respiration to form carbon dioxide, water and energy which is used for the cells' activity, but most of the glucose is converted into glycogen and stored until it is required. Some also may be converted into fat and sent to one of the body's fat depots, e.g. under the skin or around the kidneys.

Insulin (a) increases the oxidative breakdown of glucose; (b) stimulates the production of fat from glucose; (c) causes the production of glycogen from glucose; (d) inhibits the breakdown of glycogen to glucose.

Hence after a meal the blood entering the liver via the hepatic portal vein contains more glucose than the blood leaving the liver via the hepatic vein. When the absorption of food is complete the blood leaving the liver via the hepatic vein will, inversely, contain more glucose than the blood entering from the hepatic portal vein. This glucose is used, of course, by cells in cell respiration to obtain energy. Its release from the liver is accelerated by the action of adrenalin (a hormone produced by the adrenal medulla). This is particularly noticeable in times of stress when more glucose is sent out from the liver to the tissues to mobilise the body for action. The liver contains enough glycogen to supply body tissues with glucose for several hours. If the glycogen level becomes low, protein and fat can be broken down to form glycogen (a process called gluconeogenesis). This is controlled by the

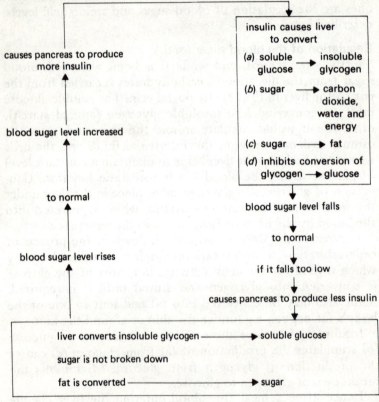

Fig. 54. *Homeostatic control of blood sugar.*

secretions from the adrenal cortex and anterior pituitary. The liver then regulates the body's blood sugar level under "instruction" from the islets of Langerhans. Production of insulin is triggered off and controlled by the blood sugar itself. If the blood sugar level is too low, less insulin is secreted and glycogen in the liver is converted into glucose and released into the blood. If it is too high, more insulin is produced and it stimulates the conversion of glucose into glycogen in the liver.

Thus the liver plays its part in the self-adjusting mechanism of homeostasis (*see* Fig. 54).

Regulation of amino acids

Amino acids in excess of the body's requirements for protein synthesis and amino acids which have come from the break-down of old cells are dealt with by the liver. They are deaminated, i.e. the NH_2 group is removed together with the hydrogen atom, to form ammonia. This is very toxic and is immediately converted into urea by a series of reactions cata-lysed by the enzyme arginase. This depends on the presence

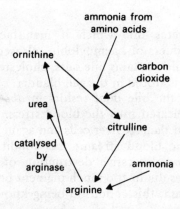

Fig. 55. *The ornithine cycle.*

of a small quantity of an amino acid called ornithine, hence the name "the ornithine cycle" (*see* Fig. 55). The ornithine combines with the ammonia and carbon dioxide to form citrulline. Citrulline unites with more ammonia to form argi-nine, which is then hydrolysed by arginase to form urea and ornithine. Urea is taken to the kidneys in the blood and excreted, while ornithine re-enters the cycle and is used again.

Regulation of lipids

The hepatic cells remove lipids from the blood and they are either broken down there or are sent to the body's fat storage depots.

Formation of cholesterol

Cholesterol is an important constituent of cell membranes; it is particularly important in nerve cells. It is a fat derivative and excess of it is secreted by the liver into the bile. If there is an excess of cholesterol relative to bile salts it can come out of solution and form gallstones which obstruct the bile duct. The cholesterol content of the diet varies, and a high cholesterol diet increases the cholesterol level in the blood. If it is not dealt with by the liver it will be deposited in the walls of the arteries thus impeding the smooth flow of blood. If it is deposited in the coronary vessels supplying the walls of the heart it can cause a clot which may result in coronary thrombosis.

Bile production

The liver excretes bile which it manufactures from the breakdown products of haemoglobin released from worn-out erythrocytes. Bile contains bile salts, cholesterol, lecithin and bile pigment; it is stored in the gall bladder.

Blockage of the bile duct results in *obstructive jaundice* when bile is released into the blood stream. Toxins, viruses and poisons can damage liver cells and again cause the release of bile into the blood stream; this condition is known as *hepatic jaundice*. Excessive destruction of erythrocytes in the liver also results in too much pigment being released into the blood stream; this condition being known as *haemolytic jaundice*.

Elimination of hormones

Sex hormones and others, e.g. A.D.H., aldosterone, are dealt with by the liver after they have fulfilled their function in the body; they are modified by the liver cells and excreted in the urine.

Formation of erythrocytes

Erythrocytes are produced in the liver of the foetus. In the adult they are formed in the bone marrow but their formation is dependent upon the presence of the haematinic principle produced in the liver from vitamin B_{12}. The body can lose its ability to absorb this vitamin since vitamin B_{12} uptake from the gut requires the intrinsic factor secreted by the gastric mucosa of the stomach. If the liver is then unable to produce the haematinic principle, the condition known as *pernicious anaemia* results. In these circumstances the average size of the erythrocytes is greater than normal and they have other abnormalities which result in their being short-lived.

The number of erythrocytes in the circulating blood drops and there is insufficient haemoglobin to carry oxygen around the body. Death can result if vitamin B_{12} or liver extract is not administered.

Elimination of haemoglobin
Old erythrocytes are phagocytosed by macrophages in the liver, the spleen and the bone marrow. The haemoglobin released in this process is split into the "haem" part and the "globin" part. The haem part is converted into biliverdin which is a green pigment; this is reduced to form brown bilirubin which is a constituent of bile and gives it colour. The iron is conserved and made into new haemoglobin. The globin is split into its constituent amino acids and these are destroyed in the liver via the ornithine cycle.

Production of heat
The liver is large, has a good blood supply and carries out many functions. It therefore has a high metabolic rate and steadily produces heat as a result of this metabolism. This raises the temperature of the blood as it passes through and thus the liver is involved in temperature regulation.

Storing blood
The liver acts as a blood reservoir. It is also a means whereby blood is transferred from the hepatic portal system to the systemic circulation. The blood supply of the liver lobules is via the sinusoids which form a very extensive spongy mesh-work between the sheets of hepatic cells. Blood enters this sinusoidal meshwork at the edge of the lobule through the interlobular branches of the portal vein and hepatic artery. Blood then passes radially through the sinusoidal spaces and drains from the lobule by the central intralobular vein. The liver contains approximately 350 cm^3 of blood but it can store up to a litre of blood, transferring it from the portal system to the systemic circulation.

Filtering the hepatic portal blood
Kupffer cells in the liver are part of the system of cells called the reticulo-endothelial system, and they engulf pieces of broken erythrocytes. Blood comes to the liver from the gut, via the hepatic portal system, and it can carry bacteria from the gut. The Kupffer cells phagocytose these bacteria in the same way as they phagocytose broken down erythrocytes. In this way the liver plays a defensive role preventing illness.

Synthesis of plasma proteins
The liver is concerned with the formation of the plasma proteins fibrinogen and albumin. Fibrinogen is responsible for the clotting of the blood. If the fibrinogen and albumin concentration in the blood falls, as it does in cases of starvation or in disease, the time taken for the blood to clot is increased in consequence.

Prothrombin is also formed in the liver, and vitamin K is essential for its synthesis. Prothrombin is the coagulant protein of blood and if it is not present, or is present in only small amounts, the blood will not clot and there is excessive bleeding. People with diets lacking vitamin K can experience uncontrolled bleeding due to the fact that the prothrombin level has fallen to a level which impairs coagulation. This can occur in the new born. Bacteria present in the gut flora of the intestines of adults synthesise vitamin K which is then absorbed through the intestinal wall into the blood. However, in order that this can occur, bile salts must be present. Obstructive jaundice means that there are too few bile salts in the gut for effective uptake of vitamin K. The deficiency of this vitamin in blood reaching the liver means that too little prothrombin is produced to bring about effective coagulation of the blood. The result is bleeding; a difficult state of affairs when an operation is required to combat the obstructive jaundice which brought about the defective uptake of vitamin K in the first place.

Storage of vitamins
The liver is concerned with the storage of several vitamins. Vitamin A is necessary to man for growth, good vision and healthy tissues in particular the tissues of the eye, (it prevents xerophthalmia). Transport of vitamin A through

the body requires an adequate supply of fat in the diet (vitamin A is a derivative of carotene the pigment present together with chlorophyll in green plants; it is a fat-soluble vitamin). Vitamin A is carried from the gut through the lymphatic system to the liver where it is stored.

Vitamin D or calciferol is a complex alcohol which:

(a) helps mobilise the calcium from the intestine so that it it can be used for building bone;

(b) is concerned with phosphate metabolism.

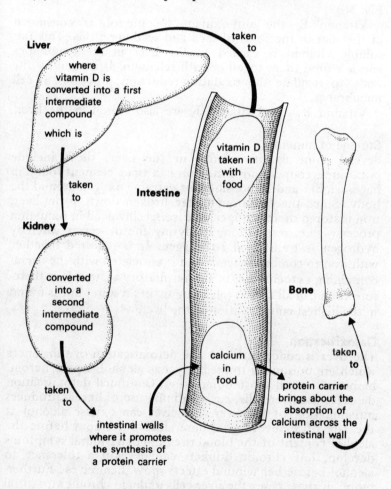

Fig. 56. *Action of vitamin D in calcium uptake.*

Vitamin D is required in the body in order that a carrier protein can be built up which will carry the calcium to the site where it is required. Vitamin D is taken in the blood to the liver where it is converted into an intermediate substance which is carried by the blood to the kidney where a second intermediate substance is formed. This second intermediate is carried in the blood to the intestine where it promotes the synthesis of the carrier protein which affects the efficient uptake of calcium from the food in the gut (*see* Fig. 56).

Vitamin E, the anti-oxidant tocopherol, is consumed in the diet in the form of seed and seed germ oils. This fat-soluble vitamin is carried to the liver in the lymphatics and is stored there together with selenium. It is used by the body to regulate the oxidation-reduction processes in cell membranes.

Vitamin B_{12} and vitamin K are also stored in the liver.

Storage of minerals

Several minerals are stored in the liver; these include potassium, copper and iron. Iron is a trace element found in haemoglobin and responsible for carrying oxygen around the body. Since the erythrocytes are broken down in the liver, iron is stored there. Copper, like iron, is involved in oxidation processes, i.e. respiration, catalysing the processes whereby hydrogen is transferred to oxygen. It is required together with iron to combat anaemia and is associated with the haem-containing cytochrome, in the respiratory system. It is therefore found in all tissues (all living tissues respire) but is found in the highest concentration in the liver.

Detoxification

The liver is concerned with the detoxification of substances which are poisonous to the body, e.g. alcohol, asprin, heroin, barbiturates and most other drugs. Continual detoxification damages the liver cells, e.g. continual use of heroin produces inflammation of the liver. The liver can oxidise alcohol at the rate of about 14 g of pure alcohol per hour before the alcohol content of the blood rises and behavioural symptoms develop, but chronic drinkers can develop a tolerance to alcohol before behavioural effects show themselves . Furthermore, in these cases the liver cells undergo chronic structural and biochemical changes in response to alcohol so that the liver cells oxidise the alcohol at a faster rate than that for normal people.

Barbiturate drugs are also dealt with by the liver; if alcohol and barbiturates are taken together the detoxification system of the liver cannot deal with them adequately and the person may die even though the barbiturate dose may be lower than the lethal dose taken on its own. Chronic alcoholism impairs the normal functioning of the liver to such an extent that it begins to accumulate fat producing a fatty liver. Further degeneration of the liver results in the hepatic cells being replaced by scar tissue, i.e. *cirrhosis* of the liver. Blood flow through the organ is impeded and there is imbalance between the plasma osmotic pressure and local blood pressure and fluid, known as ascites, accumulates in the abdominal cavity.

Alcoholics tend to neglect their diets and rely on the alcohol for their food requirements. They suffer from nutritional disorders which further aggravates the cardio-vascular disorders which develop.

This chapter should have enabled the reader to appreciate the very wide ranging functions carried out by the liver, albeit in simple outline; deeper study will be included in more advanced courses.

SELF-ASSESSMENT QUESTIONS

1. Food-laden blood is brought to the liver by the:
    (a) hepatic vein;
    (b) hepatic artery;
    (c) hepatic portal vein;
    (d) renal vein;
    (e) aorta.

2. Oxygen is brought to the liver by the:
    (a) hepatic artery;
    (b) hepatic vein;
    (c) renal artery;
    (d) hepatic portal vein;
    (e) renal vein.

3. The liver:
    (a) produces bile;
    (b) stores bile;
    (c) detoxifies bile;
    (d) adds water to bile;
    (e) concentrates bile.

4. Urea is:
   (a) produced by the pancreas;
   (b) stored by the liver;
   (c) produced by the liver;
   (d) filtered by the liver;
   (e) absorbed by the liver.

5. The liver:
   (a) stores glucose as glycogen;
   (b) deaminates glycogen;
   (c) converts glycogen to bilirubin;
   (d) converts cholesterol into glycogen;
   (e) converts glycogen into biliverdin.

6. The haematinic principle is produced in the liver from:
   (a) vitamin B_1;
   (b) vitamin B_{12};
   (c) vitamin A;
   (d) vitamin D;
   (e) vitamin E.

ASSIGNMENTS

1. Write an account of the functions of the liver, stressing its importance in the metabolism of the body.

2. Make a series of large clear diagrams to illustrate your account showing:
   (a) the position of the liver relative to the other organs in the body;
   (b) the gross anatomy of the liver;
   (c) the histological structure of the liver.

CHAPTER EIGHT

The Role of the Kidneys in Homeostasis

CHAPTER OBJECTIVES

After studying this chapter you should be able to:
* state what is meant by excretion;
* describe the gross anatomy of the urinary system of the mammal;
* identify the main parts of the mammalian kidney as seen in longitudinal section;
* make a diagram of the kidney tubule and identify its various parts;
* state the functions of the kidney;
* state the function of the kidney tubule in terms of filteration within the Bowman's capsule, reabsorption in the uriniferous tubule and secretion by the uriniferous tubules;
* explain how the kidney functions in maintaining the salt balance of the body;
* explain how the kidney maintains pH.

INTRODUCTION

Metabolism results in the production of a number of substances which are of no further use to the body. These materials are called *excretory products*. Some of these products are harmful to the body and certainly if these waste products were allowed to accumulate they would be poisonous to the body cells. The process by which the body gets rid of these waste products is called *excretion*.

Tissue respiration takes place in all the living cells within the body and involves oxidative processes resulting in hydrogen which then combines with oxygen to form water; carbon dioxide is also produced. These two excretory products are expelled from the body by the lungs.

The skin is another excretory organ which excretes a small amount of urea, water and salts as sweat.

Urea is the most abundantly formed product of protein metabolism; it is produced in the liver by deamination but is expelled from the body through the kidneys.

The liver excretes waste material from the haemoglobin of effete erythrocytes in bile.

STRUCTURE OF THE URINARY SYSTEM

The gross anatomy of the urinary system of man

The kidney is an excretory and an osmoregulatory organ. Not only does it remove the products of nitrogenous metabolism as urea but it helps to maintain the body's water content at a constant level.

The kidneys are two bean-shaped organs, approximately 12.5 cm long, lying attached to the posterior abdominal wall opposite the twelfth thoracic and first lumbar vertebrae (*see* Fig. 57).

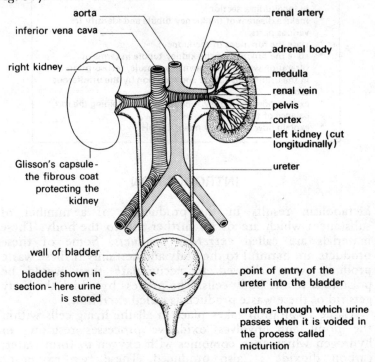

Fig. 57. *The urinary system, ventral view.*

The outer border of each kidney is convex, the inner one is concave and the depression on this side is called the hilum. The whole kidney is covered by a tough fibrous connective tissue forming Glisson's capsule. At the hilum, impure oxy-

oxygenated blood enters the kidney via the renal artery; pure deoxygenated blood leaves the kidney via the renal vein and urine leaves the kidney through the white tube called the ureter, *en route* for the bladder. The upper part of the ureter forms the pelvis of the kidney. If the kidney is cut in half longitudinally this pelvis can be seen more easily; indeed three distinct regions of kidney can be seen:

(a) the outer cortex which has numerous red dots in it — these are the glomeruli;

(b) the inner medulla region which surrounds the pelvis consists of a number of pyramid-shaped structures with their apices pointing towards the third region;

(c) the renal pelvis.

Microscopic examination of the kidney in longitudinal section
The site of urine formation is the nephron (*see* Fig. 58). There are about a million of these little tubes in the kidney. Each nephron consists of a Bowman's capsule composed of squamous epithelium and situated in the cortex of the kidney. The Bowman's capsule leads into a uriniferous tubule composed of the proximal convoluted tubule which links the capsule to the loop of Henle. The tubule continues into the distal convoluted tubule and links the loop of Henle to the collecting duct. The tubule walls are of cubical epithelium. Oxygenated blood carrying waste material is carried to the kidney via the renal artery which splits into numerous capillaries each of which goes to a Bowman's capsule where it forms a knot of capillaries called the glomerulus. Blood enters the glomerulus through the afferent arteriole and leaves it through the efferent arteriole which has a smaller bore. The blood then flows via the efferent arteriole into another capillary bed supplying the cells lining the uriniferous tubules.

THE FUNCTIONS OF THE KIDNEY

Let us list the functions of the kidney before we proceed further, and then consider each of them in detail:

(a) the excretion of urine;

(b) osmoregulation;

(c) maintenance of salt balance of the body;

(d) maintenance of the pH of the blood.

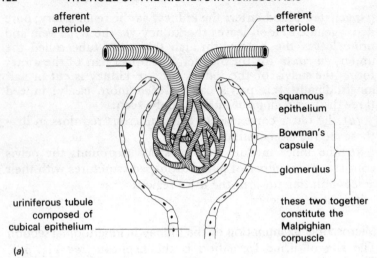

afferent
arteriole

efferent
arteriole

squamous
epithelium

Bowman's
capsule

glomerulus

uriniferous tubule
composed of
cubical epithelium

these two together
constitute the
Malpighian
corpuscle

(a)

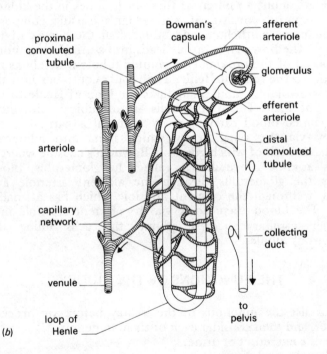

proximal
convoluted
tubule

Bowman's
capsule

afferent
arteriole

glomerulus

efferent
arteriole

distal
convoluted
tubule

arteriole

capillary
network

collecting
duct

venule

loop of
Henle

to
pelvis

(b)

Fig. 58. *(a) Malpighian corpuscle. (b) A nephron.*

How urine is formed

Urine is produced in two stages:

(a) ultrafiltration which occurs in the Bowman's capsule causing a large volume of fluid from the blood to enter the uriniferous tubules;

(b) reabsorption and secretion which occurs in the uriniferous tubules.

Filtration in the Bowman's capsule

Blood comes to the nephrons via the afferent arterioles. Pores in the glomerular walls allow small molecules to pass through them but hold back the large protein molecules and blood cells. One-fifth of the plasma filters through the capillary walls and passes into the uriniferous tubules.

The difference in bore between afferent and efferent arterioles means that the blood pressure is high enough to force the fluids through the glomerular filter. If blood pressure drops to below 70 mm of mercury (9330 Pa), urine formation ceases.

Reabsorption by the uriniferous tubules

The cubical epithelial cells lining the uriniferous tubules transport materials through their cells to the blood in the efferent arterioles coiled around the tubules. This process is called reabsorption and substances reabsorbed include sodium, glucose and amino acids, and, of course, water.

Tubular secretion by the uriniferous tubules

The cells lining the uriniferous tubules can pass substances from the blood *back* to the tubules. This is known as tubular secretion.

FLUID REGULATION

The water content of the body

Approximately 60 per cent of the body weight is made up of water. The amount varies with age and with the amount of adipose tissue present. The greater the amount of adipose tissue the less is the amount of water present.

Much of the water present in the body is extra-cellular. This includes:

(a) plasma within the vascular system containing 7 per cent of the body water;

(b) outside the vascular tissue, the lymph and interstitial fluid, making a further 18 per cent of the body water;

(c) fluids contained in cartilage and bone account for 10 per cent of the body water;

(d) transcellular fluids such as cerebrospinal fluid, aqueous humour, vitreous humour, synovial fluids, pleural, peritoneal and pericardial fluids and those in the digestive and urino-genital systems account for 2.5 per cent of the body water. The remaining 62.5 per cent of the body water is found within the cells and is therefore intracellular.

A normal man takes in approximately 2300 cm^3 of water a day in his food and drink and metabolic reactions produce another 200 cm^3; so the water balance of the body must be maintained and this is done by two processes: water intake, and water loss.

(a) Water intake is regulated by nervous mechanisms which act upon the hypothalamus creating a feeling of thirst. Intake of water then maintains the normal concentration of osmotically active substances in the body fluids.

(b) Water loss occurs in various ways:
    (i) loss through the skin as perspiration;
    (ii) loss from expired air;
    (iii) loss in faeces;
    (iv) loss in urine.

A certain minimum amount of water is required to excrete nitrogenous waste matter; when this is accomplished water loss is controlled by the body's osmoregulatory mechanism. Thus urine can be dilute and copious in amount or concentrated and small in amount.

Regulation of the sodium balance

Regulation of the sodium balance within the body indirectly determines water balance: sodium plays a major role as an osmotically active substance in the extracellular fluid; a gain in sodium means a gain in water, a loss in sodium means a loss in water. This will be considered in more detail in the next chapter.

The role of the pars nervosa

The posterior lobe of the pituitary gland, the pars nervosa, is concerned with the homeostatic regulatory mechanism for water. If this structure is damaged by disease or experimentally, the subject produces a large amount of dilute urine (as much as 40 litres per day) and compensates for the loss by drinking large amounts of water. The condition is known medically as diabetes insipidus and can be relieved by injection of pars nervosa extract (*see* Chapter 9).

Variation of the pH of the urine with diet

The pH of the urine varies with the diet of the individual. A protein-rich diet contains a high concentration of phosphates and sulphates, which affect the pH of urine making it low, i.e. the urine becomes acidic.

A diet rich in vegetable matter will be rich in potassium and other basic ions and the urine will have a high pH. Consequently human urine can vary between pH 4.8 and 7.

Regulation of the acidity of the blood by the kidney

Cells in the distal tubules actively transport hydrogen ions from the blood to the tubular fluid. In this way the kidney gets rid of excess acid from the blood via the urine.

SELF-ASSESSMENT QUESTIONS

1. Urea is a metabolic product from the breakdown of:
    (a) fats;
    (b) glucose;
    (c) alcohol;
    (d) proteins;
    (e) maltose.

2. Which one of the following blood vessels brings blood to the kidneys?
    (a) the renal vein;
    (b) the renal artery;
    (c) the pulmonary artery;
    (d) the hepatic artery;
    (e) the hepatic vein.

3. In which organ of the body is urea produced?
    (a) the pancreas;
    (b) the kidney;
    (c) the urinary bladder;
    (d) the liver;
    (e) the ureter.

4. Where does the process of blood filtration under high pressure occur?
   (a) the collecting tubule;
   (b) the pelvis of the kidney;
   (c) the uriniferous tubule;
   (d) the glomerulus;
   (e) the loop of Henle.

5. Urine is finally passed out of the body through:
   (a) the urethra;
   (b) the ureter;
   (c) the bile duct;
   (d) the uterus;
   (e) the rectum.

6. In each human kidney there are approximately how many nephrons?
   (a) 500;
   (b) 50;
   (c) 1 000 000;
   (d) 1000;
   (e) 10 000.

7. Excess amino acids are deaminated because:
   (a) they cannot be stored;
   (b) they are poisonous;
   (c) they cannot be excreted unchanged;
   (d) they upset the osmotic balance of the cells;
   (e) they are soluble.

ASSIGNMENTS

1. Write an illustrated account to show how the kidney:
   (a) acts as an osmoregulatory organ;
   (b) acts as an excretory organ.

2. A glucose molecule and a molecule of urea enter the kidney through the renal artery. Trace the routes that these two molecules may take.

How Hormones Affect Urine Production

THE BALANCE OF WATER IN BODY FLUIDS

The body gains and loses water continually; water is taken in with food and drink, it is also produced as a result of the body's metabolism, it is lost through the skin as sweat and is given out from the lungs in breath. These losses are not under the control of the body's water regulating processes. The kidneys also eliminate water in the urine and they are very responsive to the body's water requirements; they alter the amount of water lost by the body and regulate the osmotic pressure of the body fluids, i.e. carry out osmoregulation.

Urine must contain a minimum amount of water to carry the waste substances. However, urine can vary very much in its composition; the kidneys can produce large amounts of dilute urine or small amounts of concentrated urine. The factors largely responsible for the water content of urine are the sodium balance and the water balance of the tissues.

The sodium balance and water balance
The water balance is determined indirectly by the sodium balance. Sodium is the major osmotically active substance in the body's extracellular fluid. Thus a gain in sodium is followed by a gain in water and a loss in sodium is followed by

a loss in water. These important substances in the body's metabolism are controlled by two different regulatory mechanisms: thirst and antidiuretic hormone (A.D.H.) production. Sometimes these two mechanisms are activated simultaneously, e.g. when there is a severe loss of blood and therefore a need to conserve the normal composition of the extracellular fluid.

The thirst mechanism

Loss of water or increased retention of sodium results in the sensation of thirst; salivary secretion is decreased and the mouth feels dry. When water is taken in salivary secretion is increased and the concentration of body fluids returns to normal.

Osmoreceptors

The hypothalamus contains specialised nerve cells or osmoreceptors which are sensitive to the osmotic pressure of the liquids surrounding them.

Antidiuretic hormone production

The antidiuretic hormone is called vasopressin. It is produced by specialised nerve cells in the hypothalamus (*see* Fig. 59) at the base of the brain in the region known as the pars nervosa. The hormone migrates along the fibres connecting the cells to the posterior pituitary gland. Here it is stored and released into the blood.

The functions of the antidiuretic hormone

(*a*) If there is too little water in the body the osmotic pressure of the body fluids increases. The osmoreceptors respond to this and there is an increased release of A.D.H. The increased A.D.H. level in the blood causes conservation of water by most of the water in the uriniferous tubules being reabsorbed and not passing into the urine. The result is that only a little concentrated urine is produced.

(*b*) If there is too much water in the body the osmotic pressure of the body fluids decreases, less A.D.H. is produced and some of the water in the tubular urine that would have been reabsorbed continues into the urine. There is a dramatic increase in urine output.

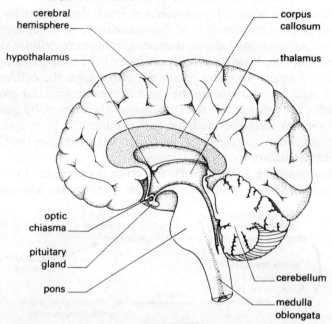

Fig. 59. *Section through the human brain showing the position of the hypothalamus.*

How A.D.H. acts on the tubule

Fluid passing from Bowman's capsule to the ureter must pass twice through the medulla, the first time through the loop of Henle, the second through the collecting duct.

In the loops, sodium is pumped out of the urine into the extracellular spaces of the medulla where it collects and becomes four times more concentrated than it is in other body fluids. The tubular walls of this section are impermeable to water which therefore cannot pass out into the tissues, so the fluid in the distal tubules becomes dilute, having lost most of its solute in the first part of the tube.

(a) If there is little A.D.H. in the blood the distal tubule and collecting duct walls remain impermeable to water. Some solutes (e.g. sugars) may be reabsorbed from these tubules but water is not. The result is very dilute urine.

(b) If there is much A.D.H. in the blood the A.D.H. acts on the distal tubules and collecting duct making them more permeable to water. Thus water can leave these tubes by osmosis and the remaining fluid then passes on into the collecting ducts. As mentioned above, the surrounding extracellular fluid is concentrated owing to the presence of the sodium pumped out of the urine so water passes out into it from the collecting duct until its contents are the same as the extracellular tissue, making the urine more concentrated, so the kidney, by pumping sodium into the medulla, can regulate the body's water content. Urine can never get more concentrated than the fluid in the medullary extracellular spaces.

Summarising then: A.D.H. secretion is determined by the osmotic pressure of the tissue fluid surrounding the osmo-

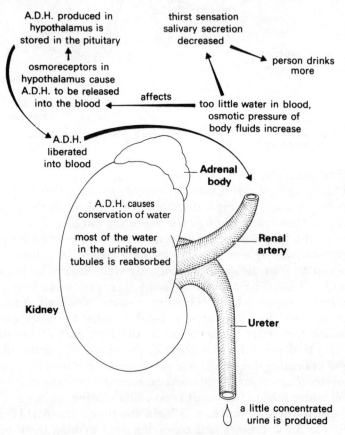

Fig. 60. *Action of A.D.H.: too little water in body fluids*

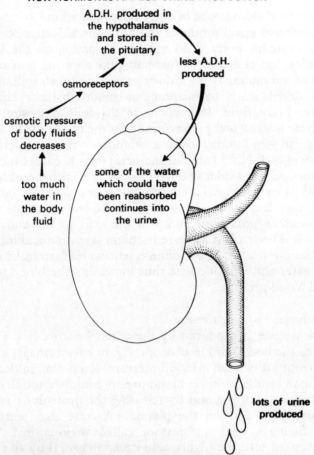

A.D.H. produced in
the hypothalamus
and stored in
the pituitary

less A.D.H.
produced

osmoreceptors

osmotic pressure
of body fluids
decreases

too much
water in
the body
fluid

some of the water
which could have
been reabsorbed
continues into
the urine

lots of urine
produced

Fig. 61. *Action of A.D.H.: too much water in body fluids.*

receptors in the hypothalamus. If the osmotic pressure surrounding the osmoreceptors in the hypothalamus rises, A.D.H. is released into the blood; it acts directly on the kidney tubules, causing them to reabsorb water. The result is that *(a)* the urine becomes more concentrated; and *(b)* there is less urine. The increased osmotic pressure also affects the thirst centre in the hypothalamus and the receptors in the stomach, with the result of a feeling of thirst so the person drinks more and the water balance in the tissues is restored (*see* Figs. 60 and 61).

The role of aldosterone in body fluid regulation

The adrenal gland produces the hormone aldosterone, which also regulates body fluid volume by acting on the kidney tubules, but this time by increasing the absorption of sodium out of the tubule. If the kidney tubules reabsorb sodium they will absorb water to maintain an osmotic balance; this conserves body fluid. It happens in the following way. Aldosterone is absorbed preferentially by the nuclei of the kidney cells. In the kidney cells it stimulates the production of messenger R.N.A. (ribonucleic acid) which causes increased formation of oxidative enzymes in the mitochondria. The result is increased A.T.P. (adenosine triphosphate) production and therefore increased availability of energy at the site of the sodium/potassium pump in the cells of the convoluted tubule. The result is increased sodium retention and increased potassium release. The sodium retention leads to reabsorption of water and chloride ions thus increasing the blood volume and blood-pressure.

Production of aldosterone

Aldosterone is produced by the adrenal cortex (*see* Fig. 62) when the body fluid level drops, e.g. in haemorrhage, low cardiac output or a fall in blood-pressure. Reduction in the blood volume results in lower blood-pressure in the arterioles in the kidneys which respond by releasing the proteolytic enzyme, *renin*. This acts on the plasma substrate angiotensinogen, producing a weak vasopressor called angiotensin I. This is converted into the highly active angiotensin II by an enzyme in the plasma. Angiotensin II is important in the maintenance of arterial pressure and may be produced continually. It is a vasoconstrictor and it stimulates the release of aldosterone from the adrenal cortex.

Summarising, then, aldosterone output responds to:
(*a*) the amount of potassium in the body fluid;
(*b*) the amount of sodium taken in with food;
(*c*) the plasma volume.

If any or all of these are low more aldosterone is liberated. It is sensed by the sensory detecting cells near the glomeruli (the juxtaglomerular cells) which cause the enzyme renin to be secreted by the kidney. This acts on angiotensinogen made in the liver and converts it into angiotensin I which is converted to angiotensin II in the plasma. It stimulates the adrenal cortex to make more aldosterone which acts on the kidneys causing sodium to be exchanged for potassium which is then excreted. Aldosterone also reduces the ratio of sodium:potassium in sweat, the fluid in the colon and also the saliva, thus it helps to retain sodium.

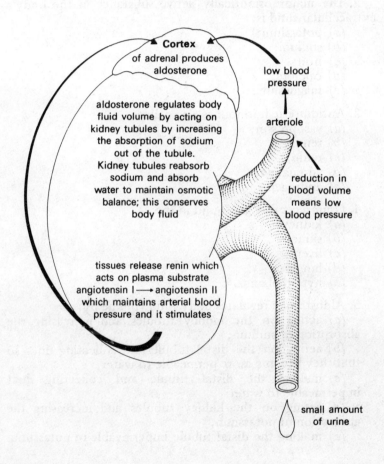

Cortex
of adrenal produces aldosterone

low blood pressure

aldosterone regulates body fluid volume by acting on kidney tubules by increasing the absorption of sodium out of the tubule.
Kidney tubules reabsorb sodium and absorb water to maintain osmotic balance; this conserves body fluid

arteriole

reduction in blood volume means low blood pressure

tissues release renin which acts on plasma substrate angiotensin I ⟶ angiotensin II which maintains arterial blood pressure and it stimulates

small amount of urine

Fig. 62. *Action of aldosterone.*

SELF-ASSESSMENT QUESTIONS

1. The principal osmoregulator is:
   (a) the skin;
   (b) the lungs;
   (c) the kidneys;
   (d) the bladder;
   (e) the osmoreceptors.

2. The major osmotically active substance in the body's extracellular fluid is:
   (a) potassium;
   (b) sodium;
   (c) iron;
   (d) calcium;
   (e) magnesium.

3. Antidiuretic hormone is also called:
   (a) vasopressin;
   (b) rennin;
   (c) renin;
   (d) adrenaline;
   (e) androgen.

4. Osmoreceptors are found in the:
   (a) kidney;
   (b) skin;
   (c) liver;
   (d) lungs;
   (e) hypothalamus.

5. Aldosterone regulates body fluid volume by:
   (a) acting on the kidney tubules and increasing the absorption of sodium;
   (b) acting on the distal tubule and collecting duct so that they become more permeable to water;
   (c) making the distal tubule and collecting duct impermeable to water;
   (d) acting on the kidney tubules and increasing the absorption of potassium;
   (e) making the distal tubule impermeable to potassium.

ASSIGNMENTS

1. Show how the hormones A.D.H. and aldosterone affect the composition of the body fluids.

2. If you were lost in the Sahara Desert without any water supply what physiological mechanisms in your body would help you to survive?

CUMULATIVE QUESTIONS, CHAPTERS 6–9

1. Enzymes requiring an acid medium in which to work are found in:
   (a) the mouth;
   (b) the duodenum;
   (c) the gall bladder;
   (d) the stomach;
   (e) the pancreas.

2. Fat absorbed from the small intestine reaches the blood via the:
   (a) lacteals in the villi;
   (b) hepatic portal vein;
   (c) lymphatics draining the small intestine;
   (d) thoracic duct;
   (e) lymph nodes.

3. Urea is produced in the:
   (a) kidney;
   (b) liver;
   (c) ureter;
   (d) uterus;
   (e) rectum.

4. Coagulation of the proteins of milk is brought about by:
   (a) renin;
   (b) rennin;
   (c) ribonucleic acid;
   (d) bile;
   (e) hydrochloric acid.

5. The liver stores:
   (a) glucose;
   (b) glycogen;
   (c) glycine;
   (d) gluten;
   (e) glycerol.

Appendix

ANSWERS TO SELF-ASSESSMENT AND CUMULATIVE QUESTIONS

Chapter 1
1. *(e)*
2. *(c)*
3. *(b)*
4. *(b)*
5. *(a)*
6. *(b)*
7. *(a)*
8. *(b)*
9. *(b)*
10. *(b)*

Chapter 2
1. *(b)*
2. *(d)*
3. *(d)*
4. *(d)*
5. *(c)*
6. *(a)*
7. *(a)*

Chapter 3
1. *(e)*
2. *(d)*
3. *(b)*
4. *(b)*
5. *(c)*
6. *(d)*
7. *(c)*
8. *(d)*

Chapter 4
1. *(a)*
2. *(d)*
3. *(c)*
4. *(b)*
5. *(b)*
6. *(e)*
7. *(e)*

Chapter 5
1. *(d)*
2. *(a)*
3. *(b)*
4. *(e)*
5. *(d)*
6. *(c)*
7. *(a)*

Cumulative questions, Chapters 1—5
1. *(a)*
2. *(d)*
3. *(c)*
4. *(d)*
5. *(e)*
6. *(d)*
7. *(d)*
8. *(c)*
9. *(c)*

Chapter 6

1. *(b)*
2. *(a)*
3. *(e)*
4. *(b)*
5. *(b)*
6. *(d)*
7. *(c)*
8. *(d)*

Chapter 7

1. *(c)*
2. *(a)*
3. *(a)*
4. *(c)*
5. *(a)*
6. *(b)*

Chapter 8

1. *(d)*
2. *(b)*
3. *(d)*
4. *(d)*
5. *(a)*
6. *(c)*
7. *(b)*

Chapter 9

1. *(c)*
2. *(b)*
3. *(a)*
4. *(e)*
5. *(a)*

Cumulative questions, Chapters 6—9

1. *(d)*
2. *(a)*
3. *(b)*
4. *(b)*
5. *(b)*

Index

acidophil (*see also* eosinophyll) 29
actin 9
adenosine diphosphate 66, 82
adenosine triphosphate 66, 82, 132
adipose tissue 13, 20
adrenal
    medulla 109
    cortex 132
adrenaline 109
aerobic respiration 66
age 76,
agglutinate 32
air 65
albumin 25
alcohol 116
aldosterone 112, 132, 133
aluminium 81
alveolus 2, 27, 60, 63
amino acids 31, 111
    essential 74
amyelinated nerve fibres 13
amylase 87, 102
anaemia 81
angiotensin I 132
angiotensis II 132, 133
anoxia (*see also* hypoxia) 65
anteaters 89
antibodies 29, 32
antidiuretic hormone (A.D.H.) 112, 128,
      129, 130, 131, 132
antigens 29, 32
anti-neuritic vitamin 78
anti-rachitic vitamin 80
antitoxins 32
anus 3
apples 79
apricots 78
aorta 39
aqueous humour 50
arachnoid 14
areolar connective tissue 19
arginase 111
arsenic 81
arteries 37, 39, 43
arteriole 37, 132
aspirin 116

atrium 38
Auerbach's plexus 99
autonomic nervous system 10
autotrophic 72
axis cylinder 11

bacon 82
bacteria 103, 104
balanced diet 71
barbiturates 116
basal metabolic rate 74, 77
basement membrane 1
basophil 29
beans 74
beer 79
beri beri 79
beverages 73
bicuspid valve (*see also* mitral valve) 39
bile 25, 50, 100, 101, 112
    canaliculi 108
bilirubin 113
biliverdin 25, 113
blackcurrants 79
black sausage 82
black treacle 82
blood 24, 113
    circulation 37, 45, 46
    clotting 30, 31, 32
    flow 45
    functions 31
    groups 32
    pressure 41, 42
    Rh factor 34
    sugar level 109
    transfusions 33
    vessel 2, 9, 43
body fluid 49
bolus 88
bone 15, 17, 82
bone marrow 28, 30, 112
boron 81
Bowman's capsule 121, 122, 123
brain 11, 12, 63, 64, 129
bread 82
breathing 64, 66
bronchiole 20, 62

139

Details of other titles
in the M&E TECbook series are to be found
on the following pages.

For a full list of titles and prices, write
for the FREE TECbook leaflet and/or the Macdonald
& Evans Technical Studies catalogue, available from
Department MP1, Macdonald & Evans Ltd.,
Estover, Plymouth PL6 7PZ

Biochemistry Level III
P.L. DAVIES
Biochemistry is a multidisciplinary subject and interacts with many traditionally understood fields of study such as physiology, cell biology, and organic and analytical chemistry. At the same time, however, it is a subject in its own right, with its own research techniques and points of emphasis. The author shows this by drawing together facts from other disciplines so that the student gradually discovers what the scope of biochemistry is and how its own special techniques are applied. The book starts by discussing subjects that are also covered by other disciplines, for example stereochemistry and energetics, and then proceeds to examine such topics as enzymes, metabolism and biochemical genetics. Although written primarily for the TEC Level III Unit, the book will also be of use to first-year undergraduates and other students who need to have a knowledge of biochemistry and of the part biochemistry plays in their own particular field of interest.
Illustrated

Electronics Level II
PETER BEARDS
This book covers the syllabus of the first TEC unit specifically concerned with electronics. It deals first with devices and then examines their applications. Semiconductor devices dominate modern electronic engineering and accordingly a study of semiconductor theory, semiconductor diodes and the junction transistor are given prominence. Thermionic valves are also described because valve theory is the basis of the cathode ray tube and valves are widely used in radio transmitters. Amplifiers (especially transistor amplifiers), waveform generation and binary logic circuits are also covered in some detail. The book is amply illustrated and includes many worked examples.
Illustrated

Multiple Choice Questions in Electrical Principles
for TEC Levels I, II and III
A. DAGGER

This book consists of 360 multiple choice questions specifically prepared for people studying or teaching Electrical or Electronic Engineering. It will be particularly useful to students who wish to use the questions for individual practice, revision or self-assessment.
Illustrated

Organisation and Procedures
in the Construction Industry
P.A. WARD

This book aims to provide the student with a clear understanding of the complex structure and workings of the construction industry. Although specifically designed to meet the requirements of the TEC Level I Unit, Organisation and Procedures, the text has been broadened to meet the needs of students participating in the joint Institute of Building/National Federation of Building Trades Employers Site Management Education and Training Scheme.
Illustrated

Accommodation Operations
COLIN DIX

This book covers the basic procedures of hotel reception departments, including reservations, billing and cashiering, and also looks at tours and groups, and the ways in which the receptionist can increase sales in a hotel. There is a chapter outlining the latest electronic systems being introduced into hotels, and discussing the direction and impact of likely future developments. The book is intended to be used by students preparing for Higher National Certificate and Diploma, TEC, HCIMA and

City and Guilds 709 Hotel Reception examinations, as well as students studying for degrees in catering management. "A useful addition for students taking exams on front office and reception." *Caterer and Hotelkeeper.*
Illustrated

ALSO AVAILABLE

Basic Biology
P.T. MARSHALL
The main theme of this HANDBOOK is the elementary physiology of the green plant and the mammal. "All that is likely to be needed for "O" Level or CSE biology is included, from the cell and the chemistry of important compounds to nutrition — both plant and mammal — respiration, co-ordination, internal control and reproduction. In short, a book well described by its title, and useful to teacher and pupil." *The Teacher*
Illustrated

Basic Botany
CLAIRE SKELLERN & PAUL ROGERS
This HANDBOOK is intended to provide a comprehensive introduction to the study of plants, with considerable emphasis being placed on the importance of applied aspects of botany. It should be useful as a comprehensive revision aid for "O" Level students, while serving as a revision source for "A" Level examinations in biology, botany, environmental studies and social biology. It will also provide background information for such studies as agriculture and horticulture.
Illustrated

Basic Organic Chemistry
W. TEMPLETON
This HANDBOOK aims to provide a concise account of the basic facts and theories of organic chemistry. It is intended to meet the requirements of students following

GCE "A" and "S" Level syllabuses, and first-year degree courses in chemistry. It will also be useful to students of medicine, pharmacy and the biological sciences. A previous knowledge of general chemistry to at least "O" Level is assumed.
Illustrated

Biology—Advanced Level
P.T. MARSHALL

This HANDBOOK is compiled along the lines of the syllabuses in Advanced Level biology operated by a number of boards. For this edition the text has been considerably revised to take account of recent findings and changes of emphasis, particularly in relation to photosynthesis, respiration, transpiration, hormones and genes, the nervous system and evolution. The examination questions at the ends of the chapters have also been revised to include those from very recent years. "The text is admirably lucid and readable." *Natural Science in Schools*
Illustrated

Chemistry for "O" Level
GEORGE USHER

This HANDBOOK contains all the essentials for students preparing for "O" Level examinations in chemistry. The chapters are short, and as far as possible each is confined to a single topic so that each section of the work can be easily understood. The text is amplified with descriptions of specific experiments and illustrated by simple diagrams of the type found in examination papers. Progress tests and specimen examination questions are included.
Illustrated

Genetics
M.W. ROBERTS

This HANDBOOK has been written for the purpose of providing a concise treatment of genetics for the student studying for "O", "A" and "S" Level examinations. As thorough a framework as is possible is provided by the book, including a simple treatment of the chemistry involved in genetics. Solved problems and a selection of sample examination questions are also included. ". . . clear and concise with good diagrams. . . highly recommended. . . ." *Natural Science in Schools*
Illustrated

Human and Social Biology
GEORGE USHER

This HANDBOOK is based on the newly revised "O" Level syllabuses now operating in a number of boards. Relevant sections have been expanded as necessary to include the material required for the Health Science syllabuses used by many overseas candidates. "This is, without doubt, an excellent and very useful little book." *Secondary Education*
Illustrated